Kelley Wingate
Math Practice

Third Grade

Credits
Content Editor: Elise Craver
Copy Editor: Angela Triplett

 Visit *carsondellosa.com* for correlations to Common Core, state, national, and Canadian provincial standards.

Carson-Dellosa Publishing, LLC
PO Box 35665
Greensboro, NC 27425 USA
carsondellosa.com

ISBN 978-1-4838-0501-6

03-085151151

Table of Contents

Introduction

Competency in basic math skills creates a foundation for the successful use of math principles in the real world. Practicing math skills—in the areas of operations, algebra, place value, fractions, measurement, and geometry—is the best way to improve at them.

This book was developed to help students practice and master basic mathematical concepts. The practice pages can be used first to assess proficiency and later as basic skill practice. The extra practice will help students advance to more challenging math work with confidence. Help students catch up, stay up, and move ahead.

Common Core State Standards (CCSS) Alignment

This book supports standards-based instruction and is aligned to the CCSS. The standards are listed at the top of each page for easy reference. To help you meet instructional, remediation, and individualization goals, consult the Common Core State Standards alignment chart on page 4.

Leveled Activities

Instructional levels in this book vary. Each area of the book offers multilevel math activities so that learning can progress naturally. There are three levels, signified by one, two, or three dots at the bottom of the page:

- Level I: These activities will offer the most support.
- Level II: Some supportive measures are built in.
- Level III: Students will understand the concepts and be able to work independently.

All children learn at their own rate. Use your own judgment for introducing concepts to children when developmentally appropriate.

Hands-On Learning

Review is an important part of learning. It helps to ensure that skills are not only covered but are internalized. The flash cards at the back of this book will offer endless opportunities for review. Use them for a basic math facts drill, or to play bingo or other fun games.

There is also a certificate template at the back of this book for use as students excel at daily assignments or when they finish a unit.

Common Core State Standards Alignment Chart

Common Core State Standards*		Practice Page(s)
Operations and Algebraic Thinking		
Represent and solve problems involving multiplication and division.	3.OA.1–3.OA.4	5–22, 24, 32–34
Understand properties of multiplication and the relationship between multiplication and division.	3.OA.5–3.OA.6	20–22, 23–25
Multiply and divide within 100.	3.OA.7	26–31
Solve problems involving the four operations, and identify and explain patterns in arithmetic.	3.OA.8–3.OA.9	32–37
Number and Operations in Base Ten		
Use place value understanding and properties of operations to perform multi-digit arithmetic.	3.NBT.1–3.NBT.3	38–46
Number and Operations—Fractions		
Develop understanding of fractions as numbers.	3.NF.1–3.NF.3	47–64
Measurement and Data		
Solve problems involving measurement and estimation of intervals of time, liquid volumes, and masses of objects.	3.MD.1–3.MD.2	65–70
Represent and interpret data.	3.MD.3–3.MD.4	71–76, 98
Geometric measurement: understand concepts of area and relate area to multiplication and to addition.	3.MD.5–3.MD.7	77–91
Geometric measurement: recognize perimeter as an attribute of plane figures and distinguish between linear and area measures.	3.MD.8	92–94
Geometry		
Reason with shapes and their attributes.	3.G.1–3.G.2	95–103

* © Copyright 2010. National Governors Association Center for Best Practices and Council of Chief State School Officers. All rights reserved.

Understanding Multiplication

To multiply means to use repeated addition. It is more easily understood if you can imagine making equal groups, and then adding all of the groups together. It looks like this:

The answer to a multiplication problem is called the **product**. The numbers being multiplied are called **factors**.

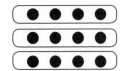

4 + 4 + 4
3 groups of 4
3 × 4 ← factors
12 ← product

Add. Then, multiply.

1.

☆☆ ☆☆
☆☆ ☆☆

_____ + _____ + _____ + _____ = _____

_____ sets of _____ equals _____

_____ × _____ = _____

2.

ⓒⓒ ⓒⓒ ⓒⓒ
ⓒⓒ ⓒⓒ ⓒⓒ

_____ + _____ + _____ = _____

_____ sets of _____ equals _____

_____ × _____ = _____

3.

_____ + _____ + _____ + _____

+ _____ = _____

_____ sets of _____ equals _____

_____ × _____ = _____

4.

☆ ☆ ☆ ☆ ☆ ☆
☆ ☆ ☆ ☆ ☆ ☆
☆ ☆ ☆ ☆ ☆ ☆

_____ + _____ + _____ + _____ + _____

+ _____ = _____

_____ sets of _____ equals _____

_____ × _____ = _____

5.

ⓒ ⓒ ⓒ
ⓒ ⓒ ⓒ

_____ + _____ = _____

_____ sets of _____ equals _____

_____ × _____ = _____

6.

_____ + _____ + _____ + _____ = _____

_____ sets of _____ equals _____

_____ × _____ = _____

Understanding Multiplication

Write an addition and multiplication problem for each picture. Then, find the sum and the product.

1.  □ + □ + □ = □ □ × □ = □	2.  □ + □ = □ □ × □ = □
3. ⊘ ⊘ ⊘ ⊘ ○ ○ ○ ○ □ + □ + □ + □ = □ □ × □ = □	4. ⊚ ⊚ ⊚ ⊚ ⊚ ⊚ ⊚ ⊚ □ + □ = □ □ × □ = □
5.  □ + □ + □ = □ □ × □ = □	6. ☆☆ ☆☆ ☆☆ ☆☆ ☆☆ ☆☆ □ + □ + □ = □ □ × □ = □
7.  □ + □ + □ = □ □ × □ = □	8. ⊚ ⊚ ⊚ ⊚ ⊚ ⊚ ⊚ ⊚ ⊚ ⊚ □ + □ = □ □ × □ = □

Understanding Multiplication

Match each multiplication problem to its addition sentence and picture. Then, solve.

1. $3 \times 3 =$ _____ $2 + 2 + 2 + 2 + 2$

2. $1 \times 5 =$ _____ $3 + 3 + 3$

3. $4 \times 2 =$ _____ 5

4. $3 \times 2 =$ _____ $4 + 4 + 4$

5. $1 \times 4 =$ _____ 4

6. $5 \times 2 =$ _____ $2 + 2 + 2 + 2$

7. $3 \times 1 =$ _____ $2 + 2 + 2$

8. $3 \times 4 =$ _____ $1 + 1 + 1$

Multiplying Sets

> To find the answer to a multiplication problem, add all of the groups together. The answer is called the **product**.
>
> Example: $3 \times 2 =$ ★★ ★★ ★★ $= 2 + 2 + 2 = 6$
>
>   (3 groups of 2)

Draw the picture. Write the multiples next to each picture. Use the picture to write an addition sentence. Then, write the multiplication sentence. The first on is done for you.

1. At the pet store, 5 dogs each have 3 spots. How many spots in all?

 3 6 9 12 15

3 + 3 + 3 + 3 + 3 = 15 spots   **5 × 3 = 15 spots**

2. On the shelf, 2 bowls each have 9 apples. How many apples total?

_____   _____

3. The secret password has 6 words that have 3 letters each. How many letters in all?

_____   _____

4. On the clothesline, 3 shirts each have 2 black stripes. How many black stripes in all?

_____   _____

5. Before they can be mailed, 7 envelopes each need 1 stamp. How many stamps are needed?

_____   _____

Name _____

Multiplying Sets

Circle groups to match the multiplication problem. Write the addition sentence. Then, find the product.

1. $4 \times 2 =$ ☆☆ ☆☆ ☆☆ ☆☆ = =

2. $3 \times 3 =$ ☆☆☆ ☆☆☆ ☆☆☆ = =

3. $3 \times 4 =$ ☆☆ ☆☆ ☆☆ ☆☆ ☆☆ ☆☆ = =

4. $2 \times 6 =$ ☆☆☆☆☆☆ ☆☆☆☆☆☆ = =

5. $2 \times 2 =$ ☆☆ ☆☆ = =

6. $2 \times 5 =$ ☆☆☆☆☆ ☆☆☆☆☆ = =

7. $4 \times 1 =$ ☆ ☆ ☆ ☆ = =

8. $4 \times 5 =$ ☆☆☆☆☆ ☆☆☆☆☆ ☆☆☆☆☆ ☆☆☆☆☆ = =

Multiplying Sets

Draw the sets. Then, write the multiplication problem. Solve.

Example: Five sets of three equals __15__. ⊙⊙⊙⊙⊙   **5 × 3 = 15**

1. Four sets of two equals _____.	2. Seven sets of three equals _____.
3. Six sets of one equals _____.	4. Five sets of four equals _____.
5. Two sets of eight equals _____.	6. Two sets of seven equals _____.
7. Nine sets of two equals _____.	8. Six sets of four equals _____.
9. One set of nine equals _____.	10. Four sets of three equals _____.
11. Eight sets of five equals _____.	12. Three sets of six equals _____.

Understanding Division

To divide means to make equal groups or to share equally. The answer to a division problem is called the **quotient**. It looks like this:

dividend   divisor   quotient

$$12 \div 3 = 4$$

Write each missing number.

1. 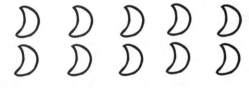 _____ ÷ _____ = _____	2.  _____ ÷ _____ = _____
3. 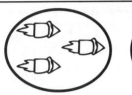 _____ ÷ _____ = _____	4. 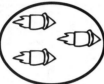 _____ ÷ _____ = _____
5. _____ ÷ _____ = _____	6. 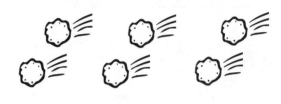 _____ ÷ _____ = _____
7. _____ ÷ _____ = _____	8.  _____ ÷ _____ = _____

Understanding Division

Circle to show a fair share. Write the division sentence. Write how many each person gets.

1.

Share with 3. _____

Each gets _____.

2.

Share with 6. _____

Each gets _____.

3.

Share with 8. _____

Each gets _____.

4.

Share with 2. _____

Each gets _____.

5.

Share with 2. _____

Each gets _____.

6.

Share with 3. _____

Each gets _____.

7.

Share with 2. _____

Each gets _____.

8.

Share with 5. _____

Each gets _____.

Understanding Division

Circle equal groups to find the quotient.

1. ☆ ☆ ☆ ☆ ☆ ☆ ☆ ☆ ☆ ☆ $10 \div 5 = \boxed{}$	2. ▢ ▢ ▢ ▢ ▢ ▢ ▢ ▢ ▢ ▢ ▢ ▢ ▢ ▢ ▢ $15 \div 3 = \boxed{}$	3. ◯ ◯ ◯ ◯ ◯ ◯ $6 \div 3 = \boxed{}$
4. ☺ ☺ ☺ ☺ ☺ ☺ ☺ ☺ $8 \div 2 = \boxed{}$	5. ☆ ☆ ☆ ☆ ☆ ☆ ☆ ☆ ☆ $9 \div 3 = \boxed{}$	6. ▢ ▢ ▢ ▢ ▢ ▢ ▢ ▢ ▢ ▢ ▢ ▢ $12 \div 4 = \boxed{}$
7. ◯ ◯ ◯ ◯ ◯ ◯ ◯ ◯ ◯ ◯ ◯ ◯ $12 \div 6 = \boxed{}$	8. ☺ ☺ ☺ ☺ ☺ ☺ ☺ ☺ ☺ ☺ ☺ ☺ ☺ ☺ ☺ ☺ ☺ ☺ $18 \div 3 = \boxed{}$	9. ☆ ☆ ☆ ☆ ☆ ☆ ☆ ☆ ☆ ☆ ☆ ☆ ☆ ☆ $14 \div 7 = \boxed{}$
10. ▢ ▢ ▢ ▢ ▢ ▢ ▢ ▢ ▢ ▢ ▢ ▢ ▢ ▢ ▢ ▢ ▢ ▢ ▢ ▢ ▢ $21 \div 3 = \boxed{}$	11. ☆ ☆ ☆ ☆ ☆ ☆ ☆ ☆ ☆ ☆ ☆ ☆ $12 \div 3 = \boxed{}$	12. ◯ ◯ ◯ ◯ ◯ ◯ ◯ ◯ ◯ ◯ ◯ ◯ ◯ ◯ ◯ ◯ $16 \div 4 = \boxed{}$

Dividing into Sets

To find the answer in a division problem, divide the total into groups with equal amounts. The answer is called the **quotient**.

Example: 6 ÷ 3 = (★★) (★★) (★★)

3 equal sets of 2 = 2

Draw the picture. Draw circles around equal sets. Use the picture to write a division sentence. The first one is done for you.

1. Dion has 12 marbles. He wants to share them with 2 friends. How many marbles do they each get?

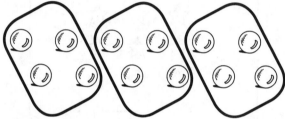

12 ÷ 3 = 4 marbles
_____

2. We picked 10 apples to make 2 pies. How many apples will be in each pie?

_____

3. Ava bought 9 flowers. How many flowers can Ava give to each of her 3 friends?

_____

4. The teacher brought 18 balls into the gym. There are 6 groups of students. How many balls should each group get?

_____

5. Luis has 7 stamps. Each letter needs 1 stamp. How many letters can Luis send?

_____

Dividing into Sets

Add circles to show the sets. Then, find the quotient.

1. $10 \div 2 =$

☆ ☆   ☆ ☆
☆   ☆
☆ ☆   ☆ ☆

$=$ ☐

2. $8 \div 2 =$

☆   ☆   ☆   ☆
☆   ☆   ☆   ☆

$=$ ☐

3. $12 \div 4 =$

☆ ☆ ☆   ☆ ☆ ☆
☆ ☆ ☆   ☆ ☆ ☆

$=$ ☐

4. $14 \div 2 =$

☆ ☆   ☆ ☆
☆ ☆ ☆   ☆ ☆ ☆
☆ ☆   ☆ ☆

$=$ ☐

5. $20 \div 5 =$

☆ ☆   ☆ ☆
☆ ☆   ☆ ☆
☆ ☆   ☆ ☆   ☆ ☆
☆ ☆   ☆ ☆   ☆ ☆

$=$ ☐

6. $9 \div 3 =$

☆   ☆   ☆
☆   ☆   ☆
☆   ☆   ☆

$=$ ☐

7. $12 \div 2 =$

☆ ☆   ☆ ☆
☆ ☆   ☆ ☆
☆ ☆   ☆ ☆

$=$ ☐

8. $18 \div 3 =$

☆ ☆ ☆ ☆ ☆ ☆
☆ ☆ ☆ ☆ ☆ ☆
☆ ☆ ☆ ☆ ☆ ☆

$=$ ☐

Dividing into Sets

Draw a picture to match the division problem. Then, write the division problem. Solve.

1. Twelve divided into three sets equals _____.

2. Eight divided into four sets equals _____.

3. Fourteen divided into two sets equals _____.

4. Twelve divided into four sets equals _____.

5. Nine divided into three sets equals _____.

6. Five divided into one set equals _____.

7. Ten divided into five sets equals _____

8. Twenty divided into five sets equals _____.

9. Twelve divided into four sets equals _____.

10. Sixteen divided into four sets equals _____.

11. Six divided into three sets equals _____.

12. Eighteen divided into six sets equals _____.

Multiplication and Division Problem Solving

Key words like **each**, **equal**, **shared**, and **evenly** can tell you that you should multiply or divide to solve the problem. But, some key words are the same for multiplication and division, like **each**.

Read the question carefully to see if you are putting together or finding a total (multiplication) or splitting apart or dividing equally (division).

Read each word problem. Decide if you should multiply or divide. Solve.

1. The Millers have 6 children. When they come to the pool, they bring 36 toys which are equally shared. How many toys does each child get?

2. We found 9 flowers. Each one had 8 petals. How many petals does that make?

3. If a car travels 40 miles per hour, how far can it travel in 3 hours?

4. Mrs. Weitz passes out 24 papers equally to 8 students. How many papers does each child get?

5. Tara and Sira share 12 cookies evenly. How many cookies does each girl get?

6. Luke is coloring eggs. Each container holds 12 eggs. Luke has 4 containers of colors. How many eggs does Luke color?

7. In the pasture there are 88 horse legs. How many horses are there?

8. We looked under 5 rocks. Each rock had 3 snails underneath it. How many snails did we see?

Multiplication and Division Problem Solving

Read each word problem. Solve. Show the multiplication or division sentence you used to find the answer.

1. On the beach are 2 towels. On each towel are 4 buckets. How many buckets are there in all?	2. Philip found 7 seashells. Each shell has an animal living in it. How many animals did Philip find?
3. Ivy invited 6 friends to her party. She wants to give each person 3 stickers. How many stickers will she need?	4. Ian has 28 grapes. He wants to put them in bags for lunch for the next 4 days. How many grapes should he put in each bag if he wants the same amount each day?
5. Our class ate 9 pizzas. Each pizza had 8 slices. How many slices did we eat in all?	6. Luke's 3 guinea pigs each ate 9 seeds. How many seeds did they eat in all?
7. A box contains 42 dog treats. Sasha divides them evenly and gives 7 treats to each of her dogs. How many dogs does Sasha have?	8. John is making lemonade for 8 people. Each glass needs 3 spoons of mix. How many spoons of lemonade mix will John use?
9. Thad ate 5 strawberries each day for a week. How many strawberries did Thad eat?	10. Yuri put 4 buttons on each doll she made. She used 32 buttons in all. How many dolls did Yuri make?

Multiplication and Division Problem Solving

Read each word problem. Solve. Show the multiplication or division sentence you used to find the answer.

1. Abigail and her 5 friends went to get ice cream. Each person got two scoops of ice cream. How many scoops did they eat in all?	2. George has to put 72 erasers into 8 boxes. How many erasers should go in each box?
3. Craig found 6 spider webs. Each web had trapped 6 bugs. How many bugs were trapped in all?	4. Darrell has 9 nickels. How much money does he have?
5. A group of 3 friends are splitting a package of 24 pencils. How many pencils should each person get?	6. Grandma has a dozen eggs. The cake recipe calls for 2 eggs. How many cakes can she bake?
7. Nina collected flowers. She put them in 8 vases. She put 9 flowers in each vase. How many flowers did she collect in all?	8. Mrs. Jacob had 40 books. She gave 8 books to each group of students. How many groups were there?
9. Reese had 56 bugs in his collection. He has 8 of each type of bug. How many types of bugs are in Reese's collection?	10. Quinn visited the library on Monday, Wednesday, and Friday. On each day, she read 6 short stories. How many stories did Quinn read in all?

11. Choose a problem above. Explain how you decided to multiply or divide.

_____

_____

_____

Unknown Numbers

Circle the correct answer in each box.

1. $\underline{\hspace{1cm}} \times 4 = 12$ Trent says 3. Kayla says 8.	2. $24 = \underline{\hspace{1cm}} \times 4$ Trent says 6. Kayla says 12.
3. $15 \div \underline{\hspace{1cm}} = 3$ Trent says 4. Kayla says 5.	4. $7 = 14 \div \underline{\hspace{1cm}}$ Trent says 2. Kayla says 7.
5. $\underline{\hspace{1cm}} \times 5 = 45$ Trent says 4. Kayla says 9.	6. $8 = \underline{\hspace{1cm}} \times 4$ Trent says 2. Kayla says 4.
7. $6 \times \underline{\hspace{1cm}} = 12$ Trent says 12. Kayla says 2.	8. $20 = 10 \times \underline{\hspace{1cm}}$ Trent says 200. Kayla says 2.
9. $30 \div \underline{\hspace{1cm}} = 10$ Trent says 3. Kayla says 300.	10. $6 = 24 \div \underline{\hspace{1cm}}$ Trent says 144. Kayla says 4.
11. $\underline{\hspace{1cm}} \div 6 = 3$ Trent says 3. Kayla says 18.	12. $8 = \underline{\hspace{1cm}} \div 4$ Trent says 32. Kayla says 6.
13. $6 \times \underline{\hspace{1cm}} = 30$ Trent says 5. Kayla says 24.	14. $36 = \underline{\hspace{1cm}} \times 6$ Trent says 20. Kayla says 6.
15. $\underline{\hspace{1cm}} \times 5 = 10$ Trent says 2. Kayla says 50.	16. $8 = \underline{\hspace{1cm}} \div 2$ Trent says 4. Kayla says 16.

Unknown Numbers

Look at the number riddles. Write the number that makes each equation true. Then, write the multiplication or division sentence that helped you solve the problem.

1. $1 \times$ ⬜ $= 6$

   ⬜ $=$ ⬜

2. 🎩 $\times 3 = 6$

   🎩 $=$ ⬜

3. $5 \times 6 =$ 🐰

   🐰 $=$ ⬜

4. ☀ $\div 8 = 1$

   ☀ $=$ ⬜

5. $14 \div$ 🌙 $= 7$

   🌙 $=$ ⬜

6. $56 \div$ ⭐ $= 7$

   ⭐ $=$ ⬜

7. $16 = 2 \times$ 🦆

   🦆 $=$ ⬜

8. $18 \div 2 =$ 🌻

   🌻 $=$ ⬜

9. $7 \times 3 = 3 \times$ 🐶

   🐶 $=$ ⬜

10. 🧶 $\div 7 = 4$

    🧶 $=$ ⬜

11. $4 \times$ 📘ABC $= 5 \times 4$

    📘ABC $=$ ⬜

12. $36 =$ ✏ 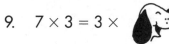 $\times 4$

    ✏ $=$ ⬜

Unknown Numbers

Find the number that makes each set of sentences true.

1. $3 \times$ _____ $= 12$, $16 \div$ _____ $= 4$

2. $16 = 4 \times$ _____, $8 \div$ _____ $= 2$

3. _____ $\times 2 = 8$, _____ $\div 1 = 4$

4. $10 =$ _____ $\times 5$, $14 \div$ _____ $= 7$

5. $6 \times 4 =$ _____, _____ $\div 3 = 8$

6. _____ $= 4 \times 9$, $6 \times 6 =$ _____

7. $6 \times$ _____ $= 18$, $3 \times$ _____ $= 9$

8. $35 = 7 \times$ _____, $45 \div$ _____ $= 9$

9. _____ $\times 4 = 32$, $16 \div$ _____ $= 2$

10. $30 =$ _____ $\times 5$, $42 \div 7 =$ _____

11. $16 \div$ _____ $= 2$, _____ $\times 7 = 56$

12. $1 = 8 \div$ _____, $64 \div$ _____ $= 8$

13. $20 \div 5 =$ _____, $36 \div$ _____ $= 9$

14. _____ $= 42 \div 6$, $28 \div 4 =$ _____

15. _____ $\div 4 = 3$, $6 \times 2 =$ _____

16. $2 =$ _____ $\div 7$, _____ $= 7 \times 2$

17. $28 \div$ _____ $= 7$, _____ $\times 5 = 20$

18. $4 = 36 \div$ _____, $3 \times 3 =$ _____

19. _____ $\div 2 = 3$, $54 \div$ _____ $= 9$

20. $5 =$ _____ $\div 3$, _____ $= 3 \times 5$

21. _____ $\times 3 = 21$, $28 \div$ _____ $= 4$

22. $72 = 9 \times$ _____, $48 \div$ _____ $= 6$

23. _____ $\div 4 = 10$, $8 \times 5 =$ _____

24. $8 =$ _____ $\div 2$, $2 \times 8 =$ _____

Connecting Multiplication and Division

Multiplication depends on equal groups, so you can use the multiplication basic facts to help you divide. The two are related, like families. They are called **fact families**. It looks like this:

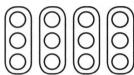

3 groups of 4

$3 \times 4 = 12$

12 divided into 4 equal groups

$12 \div 4 = 3$

Use the missing factor to help you find the quotient.

1. $2 \times \boxed{} = 8$ $8 \div 2 = \boxed{}$	2. $3 \times \boxed{} = 9$ $9 \div 3 = \boxed{}$	3. $4 \times \boxed{} = 16$ $16 \div 4 = \boxed{}$
4. $8 \times \boxed{} = 40$ $40 \div 8 = \boxed{}$	5. $5 \times \boxed{} = 25$ $25 \div 5 = \boxed{}$	6. $6 \times \boxed{} = 18$ $18 \div 6 = \boxed{}$
7. $4 \times \boxed{} = 12$ $12 \div 4 = \boxed{}$	8. $7 \times \boxed{} = 42$ $42 \div 7 = \boxed{}$	9. $3 \times \boxed{} = 15$ $15 \div 3 = \boxed{}$
10. $9 \times \boxed{} = 81$ $81 \div 9 = \boxed{}$	11. $2 \times \boxed{} = 10$ $10 \div 2 = \boxed{}$	12. $2 \times \boxed{} = 4$ $4 \div 2 = \boxed{}$
13. $5 \times \boxed{} = 20$ $20 \div 5 = \boxed{}$	14. $3 \times \boxed{} = 6$ $6 \div 3 = \boxed{}$	15. $6 \times \boxed{} = 36$ $36 \div 6 = \boxed{}$

Name _____

Connecting Multiplication and Division

Use what you know about multiplication to find each quotient. Write the related multiplication sentence. Two problems are done for you.

1. $12 \div 6 = \boxed{2}$ 2. $24 \div 4 = \boxed{}$ 3. $40 \div 5 = \boxed{}$

6 × 2 = 12

4. $16 \div 4 = \boxed{}$ 5. $21 \div 7 = \boxed{}$ 6. $9 \div 3 = \boxed{}$

7. $36 \div 6 = \boxed{}$ 8. $24 \div 8 = \boxed{}$ 9. $20 \div 4 = \boxed{}$

10. $15 \div 5 = \boxed{}$ 11. $12 \div 4 = \boxed{}$ 12. $25 \div 5 = \boxed{}$

13. $9\overline{)27}$ **3**   14. $9\overline{)36}$ 15. $9\overline{)81}$ 16. $6\overline{)54}$

9 × 3 = 27

17. $9\overline{)63}$ 18. $5\overline{)45}$ 19. $7\overline{)56}$ 20. $7\overline{)49}$

21. $8\overline{)64}$ 22. $7\overline{)42}$

© Carson-Dellosa • CD-104628

Connecting Multiplication and Division

Show two or more ways to solve each multiplication problem using properties of operations, such as using fact families or breaking it apart into easier problems.

1. $6 \times 6 =$

2. $8 \times 9 =$

3. $4 \times 7 =$

4. $7 \times 6 =$

5. $3 \times 9 =$

6. $7 \times 8 =$

7. $3 \times 8 =$

8. $4 \times 9 =$

9. $8 \times 8 =$

10. $2 \times 3 \times 3 =$

11. $4 \times 2 \times 3 =$

12. $5 \times 4 \times 2 =$

Multiplication Fluency with Factors 0–5

Solve each problem.

1. $\begin{array}{r} 5 \\ \times\, 3 \\ \hline \end{array}$
2. $\begin{array}{r} 3 \\ \times\, 4 \\ \hline \end{array}$
3. $\begin{array}{r} 5 \\ \times\, 4 \\ \hline \end{array}$
4. $\begin{array}{r} 6 \\ \times\, 3 \\ \hline \end{array}$
5. $\begin{array}{r} 1 \\ \times\, 0 \\ \hline \end{array}$
6. $\begin{array}{r} 5 \\ \times\, 6 \\ \hline \end{array}$

7. $\begin{array}{r} 9 \\ \times\, 3 \\ \hline \end{array}$
8. $\begin{array}{r} 2 \\ \times\, 0 \\ \hline \end{array}$
9. $\begin{array}{r} 2 \\ \times\, 4 \\ \hline \end{array}$
10. $\begin{array}{r} 2 \\ \times\, 3 \\ \hline \end{array}$
11. $\begin{array}{r} 8 \\ \times\, 2 \\ \hline \end{array}$
12. $\begin{array}{r} 4 \\ \times\, 8 \\ \hline \end{array}$

13. $\begin{array}{r} 3 \\ \times\, 3 \\ \hline \end{array}$
14. $\begin{array}{r} 4 \\ \times\, 3 \\ \hline \end{array}$
15. $\begin{array}{r} 4 \\ \times\, 1 \\ \hline \end{array}$
16. $\begin{array}{r} 3 \\ \times\, 0 \\ \hline \end{array}$
17. $\begin{array}{r} 5 \\ \times\, 7 \\ \hline \end{array}$
18. $\begin{array}{r} 2 \\ \times\, 0 \\ \hline \end{array}$

19. $\begin{array}{r} 2 \\ \times\, 1 \\ \hline \end{array}$
20. $\begin{array}{r} 1 \\ \times\, 7 \\ \hline \end{array}$
21. $\begin{array}{r} 9 \\ \times\, 2 \\ \hline \end{array}$
22. $\begin{array}{r} 1 \\ \times\, 0 \\ \hline \end{array}$
23. $\begin{array}{r} 4 \\ \times\, 5 \\ \hline \end{array}$
24. $\begin{array}{r} 1 \\ \times\, 4 \\ \hline \end{array}$

25. $\begin{array}{r} 8 \\ \times\, 5 \\ \hline \end{array}$
26. $\begin{array}{r} 5 \\ \times\, 2 \\ \hline \end{array}$
27. $\begin{array}{r} 5 \\ \times\, 5 \\ \hline \end{array}$

Multiplication Fluency with Factors 0–5

Solve each problem.

1. $\begin{array}{r} 2 \\ \times\, 5 \\ \hline \end{array}$ 2. $\begin{array}{r} 5 \\ \times\, 8 \\ \hline \end{array}$ 3. $\begin{array}{r} 5 \\ \times\, 3 \\ \hline \end{array}$ 4. $\begin{array}{r} 8 \\ \times\, 4 \\ \hline \end{array}$ 5. $\begin{array}{r} 3 \\ \times\, 4 \\ \hline \end{array}$ 6. $\begin{array}{r} 7 \\ \times\, 2 \\ \hline \end{array}$

7. $\begin{array}{r} 7 \\ \times\, 5 \\ \hline \end{array}$ 8. $\begin{array}{r} 1 \\ \times\, 4 \\ \hline \end{array}$ 9. $\begin{array}{r} 3 \\ \times\, 0 \\ \hline \end{array}$ 10. $\begin{array}{r} 2 \\ \times\, 2 \\ \hline \end{array}$ 11. $\begin{array}{r} 8 \\ \times\, 3 \\ \hline \end{array}$ 12. $\begin{array}{r} 4 \\ \times\, 3 \\ \hline \end{array}$

13. $\begin{array}{r} 4 \\ \times\, 6 \\ \hline \end{array}$ 14. $\begin{array}{r} 5 \\ \times\, 2 \\ \hline \end{array}$ 15. $\begin{array}{r} 4 \\ \times\, 5 \\ \hline \end{array}$ 16. $\begin{array}{r} 2 \\ \times\, 9 \\ \hline \end{array}$ 17. $\begin{array}{r} 5 \\ \times\, 5 \\ \hline \end{array}$ 18. $\begin{array}{r} 5 \\ \times\, 6 \\ \hline \end{array}$

19. $\begin{array}{r} 4 \\ \times\, 2 \\ \hline \end{array}$ 20. $\begin{array}{r} 4 \\ \times\, 9 \\ \hline \end{array}$ 21. $\begin{array}{r} 9 \\ \times\, 3 \\ \hline \end{array}$ 22. $\begin{array}{r} 4 \\ \times\, 4 \\ \hline \end{array}$ 23. $\begin{array}{r} 3 \\ \times\, 7 \\ \hline \end{array}$ 24. $\begin{array}{r} 8 \\ \times\, 2 \\ \hline \end{array}$

25. $\begin{array}{r} 6 \\ \times\, 2 \\ \hline \end{array}$ 26. $\begin{array}{r} 3 \\ \times\, 6 \\ \hline \end{array}$ 27. $\begin{array}{r} 2 \\ \times\, 0 \\ \hline \end{array}$ 28. $\begin{array}{r} 4 \\ \times\, 7 \\ \hline \end{array}$ 29. $\begin{array}{r} 3 \\ \times\, 2 \\ \hline \end{array}$ 30. $\begin{array}{r} 9 \\ \times\, 5 \\ \hline \end{array}$

31. $\begin{array}{r} 5 \\ \times\, 1 \\ \hline \end{array}$ 32. $\begin{array}{r} 2 \\ \times\, 3 \\ \hline \end{array}$ 33. $\begin{array}{r} 3 \\ \times\, 1 \\ \hline \end{array}$

Multiplication Fluency with Factors 0–5

Solve each problem.

1. $\begin{array}{r} 2 \\ \times 5 \\ \hline \end{array}$
2. $\begin{array}{r} 5 \\ \times 8 \\ \hline \end{array}$
3. $\begin{array}{r} 6 \\ \times 3 \\ \hline \end{array}$
4. $\begin{array}{r} 8 \\ \times 4 \\ \hline \end{array}$
5. $\begin{array}{r} 3 \\ \times 4 \\ \hline \end{array}$
6. $\begin{array}{r} 7 \\ \times 2 \\ \hline \end{array}$

7. $\begin{array}{r} 7 \\ \times 5 \\ \hline \end{array}$
8. $\begin{array}{r} 1 \\ \times 4 \\ \hline \end{array}$
9. $\begin{array}{r} 3 \\ \times 5 \\ \hline \end{array}$
10. $\begin{array}{r} 2 \\ \times 2 \\ \hline \end{array}$
11. $\begin{array}{r} 8 \\ \times 3 \\ \hline \end{array}$
12. $\begin{array}{r} 4 \\ \times 3 \\ \hline \end{array}$

13. $\begin{array}{r} 4 \\ \times 6 \\ \hline \end{array}$
14. $\begin{array}{r} 5 \\ \times 2 \\ \hline \end{array}$
15. $\begin{array}{r} 4 \\ \times 5 \\ \hline \end{array}$
16. $\begin{array}{r} 2 \\ \times 9 \\ \hline \end{array}$
17. $\begin{array}{r} 5 \\ \times 5 \\ \hline \end{array}$
18. $\begin{array}{r} 5 \\ \times 6 \\ \hline \end{array}$

19. $\begin{array}{r} 4 \\ \times 2 \\ \hline \end{array}$
20. $\begin{array}{r} 0 \\ \times 9 \\ \hline \end{array}$
21. $\begin{array}{r} 9 \\ \times 3 \\ \hline \end{array}$
22. $\begin{array}{r} 4 \\ \times 4 \\ \hline \end{array}$
23. $\begin{array}{r} 3 \\ \times 7 \\ \hline \end{array}$
24. $\begin{array}{r} 8 \\ \times 2 \\ \hline \end{array}$

25. $\begin{array}{r} 6 \\ \times 2 \\ \hline \end{array}$
26. $\begin{array}{r} 3 \\ \times 6 \\ \hline \end{array}$
27. $\begin{array}{r} 5 \\ \times 4 \\ \hline \end{array}$
28. $\begin{array}{r} 4 \\ \times 7 \\ \hline \end{array}$
29. $\begin{array}{r} 3 \\ \times 2 \\ \hline \end{array}$
30. $\begin{array}{r} 9 \\ \times 5 \\ \hline \end{array}$

31. $\begin{array}{r} 5 \\ \times 1 \\ \hline \end{array}$
32. $\begin{array}{r} 2 \\ \times 0 \\ \hline \end{array}$
33. $\begin{array}{r} 3 \\ \times 1 \\ \hline \end{array}$
34. $\begin{array}{r} 4 \\ \times 9 \\ \hline \end{array}$
35. $\begin{array}{r} 5 \\ \times 0 \\ \hline \end{array}$
36. $\begin{array}{r} 3 \\ \times 8 \\ \hline \end{array}$

Multiplication and Division Fluency

Solve each problem.

1. $\begin{array}{r} 8 \\ \times\, 5 \\ \hline \end{array}$ 2. $\begin{array}{r} 6 \\ \times\, 4 \\ \hline \end{array}$ 3. $\begin{array}{r} 5 \\ \times\, 5 \\ \hline \end{array}$ 4. $\begin{array}{r} 9 \\ \times\, 5 \\ \hline \end{array}$ 5. $\begin{array}{r} 4 \\ \times\, 4 \\ \hline \end{array}$ 6. $\begin{array}{r} 6 \\ \times\, 3 \\ \hline \end{array}$

7. $\begin{array}{r} 7 \\ \times\, 4 \\ \hline \end{array}$ 8. $\begin{array}{r} 7 \\ \times\, 3 \\ \hline \end{array}$ 9. $\begin{array}{r} 3 \\ \times\, 8 \\ \hline \end{array}$ 10. $\begin{array}{r} 6 \\ \times\, 2 \\ \hline \end{array}$ 11. $\begin{array}{r} 9 \\ \times\, 3 \\ \hline \end{array}$ 12. $\begin{array}{r} 5 \\ \times\, 3 \\ \hline \end{array}$

13. $\begin{array}{r} 5 \\ \times\, 4 \\ \hline \end{array}$ 14. $\begin{array}{r} 8 \\ \times\, 8 \\ \hline \end{array}$ 15. $\begin{array}{r} 5 \\ \times\, 6 \\ \hline \end{array}$ 16. $\begin{array}{r} 7 \\ \times\, 6 \\ \hline \end{array}$ 17. $\begin{array}{r} 9 \\ \times\, 8 \\ \hline \end{array}$ 18. $\begin{array}{r} 9 \\ \times\, 7 \\ \hline \end{array}$

19. $5\overline{)15}$ 20. $1\overline{)9}$ 21. $3\overline{)27}$ 22. $8\overline{)64}$ 23. $6\overline{)24}$ 24. $6\overline{)36}$

25. $8\overline{)32}$ 26. $4\overline{)36}$ 27. $8\overline{)24}$ 28. $7\overline{)14}$ 29. $2\overline{)18}$ 30. $8\overline{)8}$

Multiplication and Division Fluency

Solve each problem.

1. $\begin{array}{r} 2 \\ \times 7 \\ \hline \end{array}$
2. $\begin{array}{r} 7 \\ \times 5 \\ \hline \end{array}$
3. $\begin{array}{r} 5 \\ \times 6 \\ \hline \end{array}$
4. $\begin{array}{r} 2 \\ \times 7 \\ \hline \end{array}$
5. $\begin{array}{r} 4 \\ \times 8 \\ \hline \end{array}$
6. $\begin{array}{r} 7 \\ \times 6 \\ \hline \end{array}$

7. $\begin{array}{r} 4 \\ \times 9 \\ \hline \end{array}$
8. $\begin{array}{r} 7 \\ \times 3 \\ \hline \end{array}$
9. $\begin{array}{r} 6 \\ \times 6 \\ \hline \end{array}$
10. $\begin{array}{r} 5 \\ \times 9 \\ \hline \end{array}$
11. $\begin{array}{r} 4 \\ \times 2 \\ \hline \end{array}$
12. $\begin{array}{r} 2 \\ \times 6 \\ \hline \end{array}$

13. $\begin{array}{r} 3 \\ \times 6 \\ \hline \end{array}$
14. $\begin{array}{r} 8 \\ \times 7 \\ \hline \end{array}$
15. $\begin{array}{r} 8 \\ \times 6 \\ \hline \end{array}$
16. $\begin{array}{r} 5 \\ \times 7 \\ \hline \end{array}$
17. $\begin{array}{r} 1 \\ \times 5 \\ \hline \end{array}$
18. $\begin{array}{r} 4 \\ \times 5 \\ \hline \end{array}$

19. $\begin{array}{r} 8 \\ \times 5 \\ \hline \end{array}$
20. $\begin{array}{r} 4 \\ \times 3 \\ \hline \end{array}$
21. $\begin{array}{r} 6 \\ \times 5 \\ \hline \end{array}$
22. $\begin{array}{r} 7 \\ \times 7 \\ \hline \end{array}$
23. $\begin{array}{r} 9 \\ \times 6 \\ \hline \end{array}$
24. $\begin{array}{r} 7 \\ \times 9 \\ \hline \end{array}$

25. $8\overline{)64}$
26. $9\overline{)18}$
27. $6\overline{)48}$
28. $6\overline{)54}$
29. $8\overline{)56}$
30. $4\overline{)32}$

31. $3\overline{)24}$
32. $2\overline{)14}$
33. $8\overline{)16}$
34. $5\overline{)45}$
35. $4\overline{)20}$
36. $9\overline{)81}$

Multiplication and Division Fluency

Solve each problem.

1. $\begin{array}{r} 9 \\ \times\,4 \\ \hline \end{array}$
2. $\begin{array}{r} 6 \\ \times\,9 \\ \hline \end{array}$
3. $\begin{array}{r} 5 \\ \times\,6 \\ \hline \end{array}$
4. $\begin{array}{r} 5 \\ \times\,3 \\ \hline \end{array}$
5. $\begin{array}{r} 7 \\ \times\,9 \\ \hline \end{array}$
6. $\begin{array}{r} 8 \\ \times\,3 \\ \hline \end{array}$

7. $\begin{array}{r} 6 \\ \times\,7 \\ \hline \end{array}$
8. $\begin{array}{r} 4 \\ \times\,5 \\ \hline \end{array}$
9. $\begin{array}{r} 4 \\ \times\,2 \\ \hline \end{array}$
10. $\begin{array}{r} 4 \\ \times\,4 \\ \hline \end{array}$
11. $\begin{array}{r} 9 \\ \times\,7 \\ \hline \end{array}$
12. $\begin{array}{r} 8 \\ \times\,8 \\ \hline \end{array}$

13. $\begin{array}{r} 7 \\ \times\,2 \\ \hline \end{array}$
14. $\begin{array}{r} 8 \\ \times\,6 \\ \hline \end{array}$
15. $\begin{array}{r} 6 \\ \times\,8 \\ \hline \end{array}$
16. $\begin{array}{r} 4 \\ \times\,6 \\ \hline \end{array}$
17. $\begin{array}{r} 8 \\ \times\,4 \\ \hline \end{array}$
18. $\begin{array}{r} 6 \\ \times\,3 \\ \hline \end{array}$

19. $\begin{array}{r} 8 \\ \times\,5 \\ \hline \end{array}$
20. $\begin{array}{r} 9 \\ \times\,8 \\ \hline \end{array}$
21. $\begin{array}{r} 4 \\ \times\,3 \\ \hline \end{array}$
22. $\begin{array}{r} 8 \\ \times\,9 \\ \hline \end{array}$
23. $\begin{array}{r} 6 \\ \times\,5 \\ \hline \end{array}$
24. $\begin{array}{r} 9 \\ \times\,2 \\ \hline \end{array}$

25. $5\overline{)30}$
26. $4\overline{)36}$
27. $2\overline{)18}$
28. $4\overline{)36}$
29. $3\overline{)27}$
30. $9\overline{)9}$

31. $3\overline{)24}$
32. $4\overline{)32}$
33. $3\overline{)9}$
34. $8\overline{)56}$
35. $4\overline{)36}$
36. $8\overline{)32}$

37. $8\overline{)64}$
38. $9\overline{)81}$
39. $7\overline{)28}$
40. $7\overline{)49}$
41. $8\overline{)16}$
42. $1\overline{)4}$

Multistep Word Problems

Anthony and his Scout troop hiked 2 miles and then rested. After their break, they hiked another 3 miles. The total distance of the hike was 7 miles. How many more miles did they need to go before reaching their destination?

First: Add the total distance hiked so far. $2 + 3 = 5$

Then: Subtract this sum from the total to find the remaining distance. $7 - 5 = 2$

Answer: They must hike 2 more miles to reach their destination.

Read each problem. Show your work for both parts of the problem. Record your answer in the space given.

1. Our family drove 453 miles on vacation. We crossed 4 states. We stopped 2 times in 3 states and 3 times in the last state. How many times did we stop? First: _____ Then: _____ We stopped _____ times.	2. We went to an amusement park. There were 4 roller coasters. Jill rode each one 3 times. Henry rode each one twice. How many times did Jill and Henry ride them in all? First: _____ Then: _____ They rode _____ times.
3. We hiked 9 miles on both Monday and Tuesday. We rode our bikes 23 miles on Tuesday and 31 miles on Wednesday. How far did we travel altogether? First: _____ Then: _____ We traveled _____ miles.	4. Mia reads books in the car. On one trip, she read 173 pages. On another trip, she read 194 pages. If the book is 421 pages long, how many pages does she have left to read? First: _____ Then: _____ She has _____ pages left.
5. Jared's goal is to find 1,000 insects. In one park, he counted 356 bugs. In another park, he counted 358 bugs. How many more does he need to find? First: _____ Then: _____ He needs to find _____ more bugs.	6. Laura found 23 flowers in her backyard. She found 22 more in her friend's yard. She divided the flowers equally between 5 vases. How many flowers did she put in each vase? First: _____ Then: _____ She put _____ flowers in each vase.

Multistep Word Problems

Use two or more steps to solve each problem.

1. Lola and Mariah have each blown up 4 balloons. They need a total of 20 balloons for the party at school. How many more do they need to blow up?

    $4 \times 2 = \boxed{}$

    then, $20 - \boxed{} = \boxed{}$ They need $\boxed{}$ more balloons.

2. Lee has collected a total of 29 cans and boxes of food for the food drive. Chung has collected 27. Their team goal was to collect 75 cans and boxes of food altogether. How many more do they need to collect to reach their awesome goal?

3. Yesterday, Eric picked 34 apples off the tree in his backyard. Today, he picked another 66 apples. His father asked him to divide the total in half so that they could share their apples with the food bank. How many apples did they donate?

4. Roberto, Heather, and Macon each raised 3 dollars for the fund-raiser at school by selling pretzels at the carnival. They need a total of 15 dollars to reach their goal. How much more do they need to raise to reach their goal?

5. Wyatt pays 2 dollars each time he goes to the water park to swim. He has already been there 5 times this summer! How many more times can he go before he spends a total of 20 dollars?

6. Zaina loves to draw! She draws on 8 sheets of her pad a day. She's had her 80-page drawing pad for 9 days. How many clean pages does she have left?

Multistep Word Problems

Solve. Show your work.

1. Seth has 3 nickels in each of his 4 pockets. How much money does Seth have?	2. Three bags each have 9 marbles. Ten of the marbles rolled under the table. How many are left?
3. Four bags each hold 8 fruit snacks. Five snacks were eaten. Are there enough left to share equally with 3 people? Explain.	4. Four cups are on the table. Each cup needs 6 gumdrops in it. One bag has 25 gumdrops in it. Is one bag enough to fill the 4 cups? Explain.
5. Victor practiced his free throws. On 3 days, he made 9 shots each day. On 4 days, he made 7 shots each day. How many shots did he make all week?	6. Six words each have 2 vowels and 2 consonants. How many letters are there in the 6 words? How many vowels? Consonants?
7. Seven bags each hold 9 unit cubes. Twenty-one more cubes were found. What is the total number of cubes? If those 21 cubes were put evenly into the original 7 bags, how many unit cubes would there be in each bag?	8. Pilar used 253 beads to make a bracelet. She used 704 beads to make a necklace. If the package started with 1,000 beads, how many beads are left?

Arithmetic Patterns

Write the products to complete the table.

×	0	1	2	3	4	5	6	7	8	9	10
0											
1											
2											
3											
4											
5											
6											
7											
8											
9											
10											

1. Color each product of double factors red.

2. Complete each sentence.

    Multiples of 0 always equal _____.

    Multiples of 10 always end with _____.

    Multiples of 5 always end with _____ or _____.

3. Write *even* or *odd* in each blank.

    even number × even number = _____ number

    odd number × odd number = _____ number

    even number × odd number = _____ number

Arithmetic Patterns

Fill in the answers on the times table below.

$\times$	0	1	2	3	4	5	6	7	8	9	10
0											
1											
2											
3											
4											
5											
6											
7											
8											
9											
10											

Circle *odd*, *even*, or *both* to describe the patterns in the table.

1.	Count by ones.	odd, even, both	6.	Count by sixes.
2.	Count by twos.	odd, even, both	7.	Count by sevens.
3.	Count by threes.	odd, even, both	8.	Count by eights.
4.	Count by fours.	odd, even, both	9.	Count by nines.
5.	Count by fives.	odd, even, both	10.	Count by tens.

1. Count by ones.   odd, even, both   6. Count by sixes.   odd, even, both
2. Count by twos.   odd, even, both   7. Count by sevens.   odd, even, both
3. Count by threes.   odd, even, both   8. Count by eights.   odd, even, both
4. Count by fours.   odd, even, both   9. Count by nines.   odd, even, both
5. Count by fives.   odd, even, both   10. Count by tens.   odd, even, both

11. Write *even* or *odd* in each blank.

even number $\times$ even number = _____ number

odd number $\times$ odd number = _____ number

even number $\times$ odd number = _____ number

Arithmetic Patterns

Write the products to complete the table.

×	0	1	2	3	4	5	6	7	8	9	10
0											
1											
2											
3											
4											
5											
6											
7											
8											
9											
10											

1. Find a pattern in the table. Describe it.

    _____

    _____.

    Why does the pattern happen? How can it help you solve other problems?

    _____

    _____.

2. Find a different pattern in the table. Describe it.

    _____

    _____.

    Why does the pattern happen? How can it help you solve other problems?

    _____

    _____.

Rounding Numbers

Rounding numbers is a way of replacing one number with another number that tells about how many or how much.

When rounding to the nearest number, look at the digit to the right of it. If that column has 0, 1, 2, 3, or 4 in it, round down. If the column has 5, 6, 7, 8, or 9 in it, round up.

Round 23 to the nearest ten. Look at the ones digit.	Round 284 to the nearest hundred. Look at the tens digit.
20 **23** 30	200 **284** 300
Round 23 down to 20.	Round 284 up to 300.

Round the numbers to the place value listed.

1. 23 _____ 567 _____ 775 _____ ten	2. 483 _____ 809 _____ 495 _____ hundred	3. 609 _____ 937 _____ 148 _____ hundred
4. 2,813 _____ 408 _____ 742 _____ ten	5. 311 _____ 407 _____ 3,054 _____ hundred	6. 384 _____ 99 _____ 826 _____ ten

Rounding Numbers

Draw an arrow to show which way to round the number to the nearest ten or hundred. Then, write the rounded number. The first one is done for you.

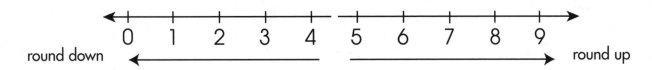

round down ⟵   round up ⟶

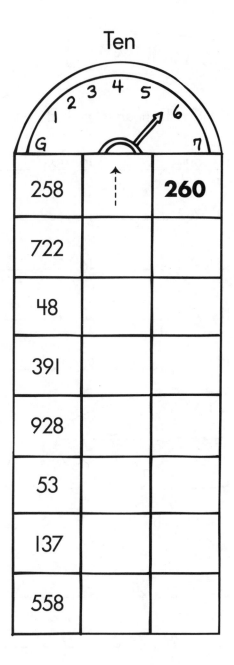

Ten

258	↑	**260**
722		
48		
391		
928		
53		
137		
558		

Hundred

902		
445		
84		
231		
616		
760		
29		
870		

Rounding Numbers

Round to the nearest ten.

1. 72 _____   2. 14 _____

3. 83 _____   4. 49 _____

5. 55 _____   6. 62 _____

7. 17 _____   8. 29 _____

9. 34 _____   10. 95 _____

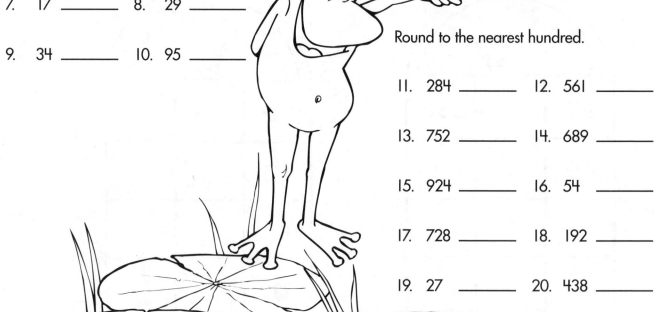

LOOK RIGHT

Round to the nearest hundred.

11. 284 _____   12. 561 _____

13. 752 _____   14. 689 _____

15. 924 _____   16. 54 _____

17. 728 _____   18. 192 _____

19. 27 _____   20. 438 _____

Round to the underlined place value. The first one is done for you.

21. 1,4<u>3</u>2   **1,430**   22. <u>4</u>18 _____   23. <u>2</u>42 _____

24. 4,<u>2</u>99 _____   25. 6,<u>4</u>19 _____   26. 7,5<u>4</u>6 _____

Addition and Subtraction within 1,000

When adding, if the sum in the tens column is greater than 9, regroup to the hundreds column. It looks like this:

1. Add the ones. Regroup if needed.
```
  172
+ 473
    5
```

2. Add the tens. Regroup if needed.
```
  1
  172
+ 473
   45
```

3. Add the hundreds.
```
  1
  172
+ 473
  645
```

When subtracting, if the top number in the tens column is less than the bottom number, you must regroup from the hundreds. It looks like this:

1. Subtract the ones. Regroup if needed.
```
  607
- 284
    3
```

2. Subtract the tens. Regroup if needed.
```
 5 10
  6̸0̸7
- 284
   23
```

3. Subtract the hundreds.
```
 5 10
  6̸0̸7
- 284
  323
```

Solve each problem. Regroup when necessary.

1. 634 + 268   2. 987 + 489   3. 768 − 479   4. 888 + 276   5. 747 − 458   6. 950 − 580

7. 394 + 496   8. 689 − 478   9. 254 + 347   10. 665 + 337   11. 521 − 295   12. 988 + 748

13. 301 − 242   14. 349 + 233   15. 878 + 287   16. 727 − 49   17. 348 + 948   18. 847 − 358

19. 477 + 298   20. 704 − 597   21. 846 − 457   22. 834 + 249   23. 405 − 228   24. 118 + 953

25. 703 − 478   26. 653 + 307   27. 584 − 295   28. 600 − 367   29. 113 + 298   30. 393 + 298

Addition and Subtraction within 1,000

Solve each problem. Regroup when necessary.

1. 784
  − 591

2. 979
  + 654

3. 945
  + 379

4. 825
  − 638

5. 870
  + 739

6. 478
  + 655

7. 675
  + 597

8. 456
  + 327

9. 654
  − 265

10. 293
  − 187

11. 765
  + 428

12. 824
  − 548

13. 845
  − 566

14. 349
  + 233

15. 434
  + 948

16. 725
  − 469

17. 827
  − 529

18. 638
  + 422

19. 539
  + 468

20. 574
  − 293

21. 955
  + 134

22. 423
  − 155

23. 588
  + 294

24. 536
  − 258

25. 666
  + 291

26. 857
  − 675

27. 646
  + 668

28. 748
  − 353

29. 831
  − 357

30. 348
  + 489

31. 239
  + 293

32. 438
  − 254

33. 487
  − 391

34. 159
  + 485

35. 648
  + 437

36. 940
  − 556

Addition and Subtraction within 1,000

Solve each problem. Regroup when necessary.

1.
$$539 - 375$$

2.
$$476 + 243$$

3.
$$176 + 484$$

4.
$$392 + 292$$

5.
$$787 - 598$$

6.
$$165 + 427$$

7.
$$481 + 428$$

8.
$$842 + 177$$

9.
$$762 - 395$$

10.
$$856 - 399$$

11.
$$347 + 983$$

12.
$$275 + 298$$

13.
$$628 - 127$$

14.
$$531 - 467$$

15.
$$496 - 288$$

16.
$$389 + 392$$

17.
$$276 + 391$$

18.
$$374 - 276$$

19.
$$983 - 468$$

20.
$$834 - 376$$

21.
$$452 + 287$$

22.
$$392 + 284$$

23.
$$597 - 387$$

24.
$$584 - 287$$

25.
$$498 - 268$$

26.
$$735 + 373$$

27.
$$395 + 161$$

28.
$$502 - 321$$

29.
$$890 - 249$$

30.
$$848 - 399$$

31.
$$400 - 373$$

32.
$$837 - 209$$

33.
$$622 - 323$$

34.
$$786 - 389$$

35.
$$663 - 261$$

36.
$$429 - 188$$

Multiplying by 10s

Follow these steps to multiply with a multiple of 10.

1. Multiply the nonzero digits.

$$\begin{array}{r} 30 \\ \times\ 4 \\ \hline 12 \end{array}$$

2. Add a 0. Because the top number you multiplied by is 10 times larger, the product is too.

$$\begin{array}{r} 30 \\ \times\ 4 \\ \hline 120 \end{array}$$

Multiply. The first one is done for you.

1. $\begin{array}{r} 90 \\ \times\ 2 \\ \hline \mathbf{180} \end{array}$

2. $\begin{array}{r} 60 \\ \times\ 3 \\ \hline \end{array}$

3. $\begin{array}{r} 80 \\ \times\ 8 \\ \hline \end{array}$

4. $\begin{array}{r} 40 \\ \times\ 3 \\ \hline \end{array}$

5. $\begin{array}{r} 60 \\ \times\ 9 \\ \hline \end{array}$

6. $\begin{array}{r} 70 \\ \times\ 2 \\ \hline \end{array}$

7. $\begin{array}{r} 90 \\ \times\ 7 \\ \hline \end{array}$

8. $\begin{array}{r} 50 \\ \times\ 4 \\ \hline \end{array}$

9. $\begin{array}{r} 70 \\ \times\ 3 \\ \hline \end{array}$

10. $\begin{array}{r} 60 \\ \times\ 6 \\ \hline \end{array}$

11. $\begin{array}{r} 50 \\ \times\ 2 \\ \hline \end{array}$

12. $\begin{array}{r} 70 \\ \times\ 5 \\ \hline \end{array}$

13. $\begin{array}{r} 20 \\ \times\ 4 \\ \hline \end{array}$

14. $\begin{array}{r} 20 \\ \times\ 3 \\ \hline \end{array}$

15. $\begin{array}{r} 30 \\ \times\ 8 \\ \hline \end{array}$

16. $\begin{array}{r} 20 \\ \times\ 5 \\ \hline \end{array}$

$$\begin{array}{r} 50 \\ \times\ 5 \\ \hline 250 \end{array}$$

Multiplying by 10s

Multiply. The first one is done for you.

1. $\begin{array}{r} 90 \\ \times\ 2 \\ \hline \mathbf{180} \end{array}$

2. $\begin{array}{r} 60 \\ \times\ 5 \\ \hline \end{array}$

3. $\begin{array}{r} 70 \\ \times\ 4 \\ \hline \end{array}$

4. $\begin{array}{r} 50 \\ \times\ 7 \\ \hline \end{array}$

5. $\begin{array}{r} 60 \\ \times\ 9 \\ \hline \end{array}$

6. $\begin{array}{r} 50 \\ \times\ 3 \\ \hline \end{array}$

7. $\begin{array}{r} 20 \\ \times\ 6 \\ \hline \end{array}$

8. $\begin{array}{r} 60 \\ \times\ 8 \\ \hline \end{array}$

9. $\begin{array}{r} 50 \\ \times\ 8 \\ \hline \end{array}$

10. $\begin{array}{r} 30 \\ \times\ 7 \\ \hline \end{array}$

11. $\begin{array}{r} 20 \\ \times\ 9 \\ \hline \end{array}$

12. $\begin{array}{r} 80 \\ \times\ 6 \\ \hline \end{array}$

13. $\begin{array}{r} 90 \\ \times\ 9 \\ \hline \end{array}$

14. $\begin{array}{r} 70 \\ \times\ 6 \\ \hline \end{array}$

15. $\begin{array}{r} 80 \\ \times\ 8 \\ \hline \end{array}$

16. $\begin{array}{r} 30 \\ \times\ 8 \\ \hline \end{array}$

Multiplying by 10s

Multiply.

1. $\begin{array}{r} 20 \\ \times\ 3 \\ \hline \end{array}$

2. $\begin{array}{r} 20 \\ \times\ 4 \\ \hline \end{array}$

3. $\begin{array}{r} 40 \\ \times\ 2 \\ \hline \end{array}$

4. $\begin{array}{r} 10 \\ \times\ 7 \\ \hline \end{array}$

5. $\begin{array}{r} 10 \\ \times\ 5 \\ \hline \end{array}$

6. $\begin{array}{r} 20 \\ \times\ 9 \\ \hline \end{array}$

7. $\begin{array}{r} 90 \\ \times\ 3 \\ \hline \end{array}$

8. $\begin{array}{r} 40 \\ \times\ 6 \\ \hline \end{array}$

9. $\begin{array}{r} 60 \\ \times\ 4 \\ \hline \end{array}$

10. $\begin{array}{r} 90 \\ \times\ 9 \\ \hline \end{array}$

11. $\begin{array}{r} 60 \\ \times\ 7 \\ \hline \end{array}$

12. $\begin{array}{r} 30 \\ \times\ 5 \\ \hline \end{array}$

13. $\begin{array}{r} 30 \\ \times\ 8 \\ \hline \end{array}$

14. $\begin{array}{r} 40 \\ \times\ 3 \\ \hline \end{array}$

15. $\begin{array}{r} 70 \\ \times\ 5 \\ \hline \end{array}$

16. $\begin{array}{r} 80 \\ \times\ 6 \\ \hline \end{array}$

17. $\begin{array}{r} 90 \\ \times\ 2 \\ \hline \end{array}$

18. $\begin{array}{r} 30 \\ \times\ 4 \\ \hline \end{array}$

19. $\begin{array}{r} 40 \\ \times\ 6 \\ \hline \end{array}$

20. $\begin{array}{r} 80 \\ \times\ 3 \\ \hline \end{array}$

21. $\begin{array}{r} 70 \\ \times\ 7 \\ \hline \end{array}$

22. $\begin{array}{r} 90 \\ \times\ 4 \\ \hline \end{array}$

23. $\begin{array}{r} 70 \\ \times\ 8 \\ \hline \end{array}$

24. $\begin{array}{r} 50 \\ \times\ 8 \\ \hline \end{array}$

Understanding Unit Fractions

This circle has been divided into equal parts, so it can also be described as a **fraction**.

Example:

■ = $\frac{1}{4}$ ← ■ parts
           ← total parts

Use the circles to answer the questions.

1.

There are _____ equal parts.

[____] of the whole is shaded.

2.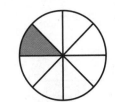

There are _____ equal parts.

[____] of the whole is shaded.

3.

There are _____ equal parts.

[____] of the whole is shaded.

4.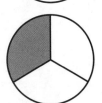

There are _____ equal parts.

[____] of the whole is shaded.

5.

There are _____ equal parts.

[____] of the whole is shaded.

6.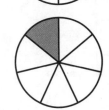

There are _____ equal parts.

[____] of the whole is shaded.

Understanding Unit Fractions

Shade and label the unit fraction for each whole. Some fractions may be used more than once.

$$\frac{1}{2} \qquad \frac{1}{3} \qquad \frac{1}{4} \qquad \frac{1}{6} \qquad \frac{1}{8}$$

1.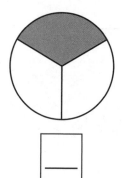

2.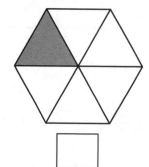

3.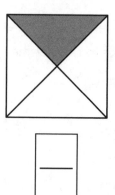

4.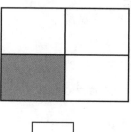

5.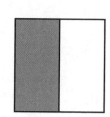

6.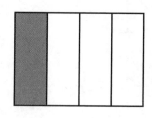

7.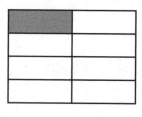

8.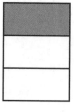

9.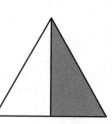

Understanding Unit Fractions

Write the unit fraction to label each part of the whole. The first one is done for you.

1.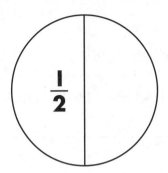

2.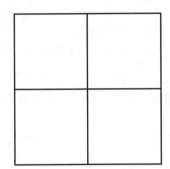

3.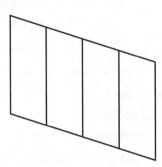

4.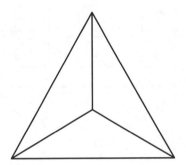

5.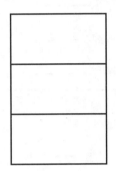

6.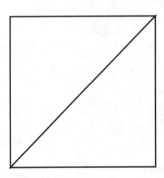

7.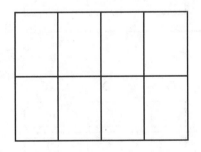

8.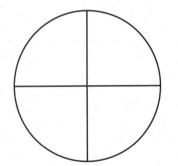

9.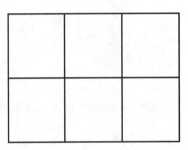

10.

11.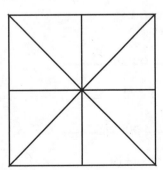

12.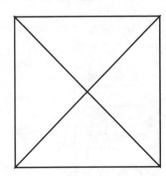

Identifying Fractions

Fractions of a whole

 $\dfrac{4}{8}$ numerator
denominator

How many parts are shaded? 4 (numerator)
How many equal parts in total? 8 (denominator)

Look at the pictures. Fill in the blanks. Then, write each fraction under the correct letter. The first one is started for you.

1. Total number of equal parts: __5__

   a. white = __2__ out of __5__

   b. dotted = _____ out of _____

   c. shaded = _____ out of _____

   a. $\dfrac{2}{5}$ b. ☐/☐   c. ☐/☐

2. Total number of equal parts: _____

   a. dotted = _____ out of _____

   b. striped = _____ out of _____

   c. stars = _____ out of _____

   a. ☐/☐   b. ☐/☐   c. ☐/☐

3. Total number of equal parts: _____

   a. hearts = _____ out of _____

   b. striped = _____ out of _____

   c. white = _____ out of _____

   a. ☐/☐   b. ☐/☐   c. ☐/☐

4. Total number of equal parts: _____

   a. hearts = _____ out of _____

   b. stars = _____ out of _____

   c. shaded = _____ out of _____

   d. striped = _____ out of _____

   e. white = _____ out of _____

   a. ☐/☐   b. ☐/☐   c. ☐/☐

   d. ☐/☐   e. ☐/☐

Identifying Fractions

Complete each fraction of the shaded part.

1. 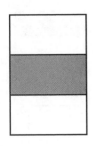 $\dfrac{}{3}$

2. 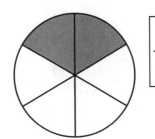 $\dfrac{2}{}$

3. 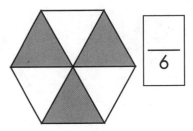 $\dfrac{}{6}$

4. 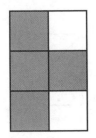 $\dfrac{}{1}$

5. 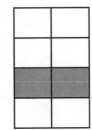 $\dfrac{3}{}$

6. 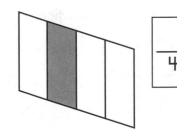 $\dfrac{}{8}$

7. 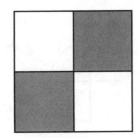 $\dfrac{2}{}$

8. 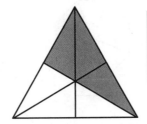 $\dfrac{}{4}$

9. 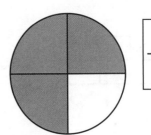 $\dfrac{}{4}$

10. 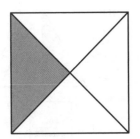 $\dfrac{}{6}$

11. 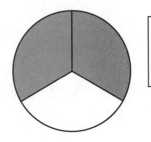 $\dfrac{5}{}$

12. 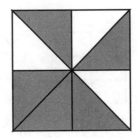 $\dfrac{}{4}$

Identifying Fractions

What is the fraction of the shaded part?

1.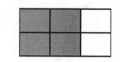

2.

3.

4.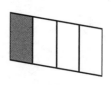

5.

6.

7.

8.

9.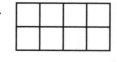

10.

Shade to show the fraction.

11.

$\dfrac{1}{4}$

12.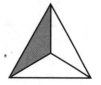

$\dfrac{2}{6}$

13.

$\dfrac{3}{4}$

14.

$\dfrac{6}{8}$

15.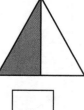

$\dfrac{5}{6}$

16.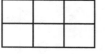

$\dfrac{4}{6}$

17.

$\dfrac{2}{3}$

18.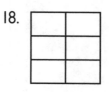

$\dfrac{1}{6}$

19.

$\dfrac{4}{8}$

20.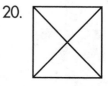

$\dfrac{2}{4}$

Understanding Fractions on a Number Line

Fractions can be represented on a number line.

Both the number line and the rectangle represent the fraction $\frac{1}{3}$.

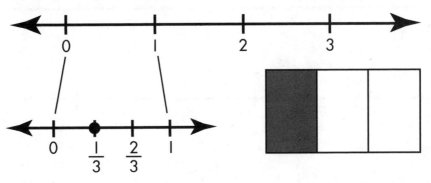

Use the number lines to answer the questions.

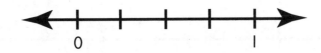

1. The number line is divided into _____ parts.

2. Label the number line with the correct fractions.

3. Each section of the number line represents what fraction? _____

4. Draw a dot to show the location of $\frac{2}{4}$.

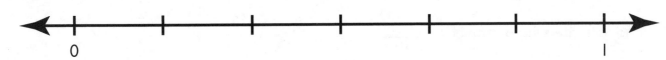

5. The number line is divided into _____ parts.

6. Label the number line with the correct fractions.

7. Each section of the number line represents what fraction? _____

8. Draw a dot to show the location of $\frac{5}{6}$.

Understanding Fractions on a Number Line

Use the number lines to answer the questions.

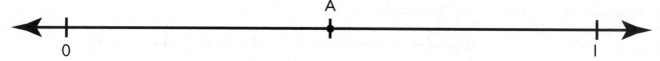

1. Label each number line with the correct fractions.

2. Which dot shows the location of $\frac{2}{3}$? _____

3. Which dot shows the location of $\frac{1}{4}$? _____

4. Which dot shows the location of $\frac{4}{6}$? _____

5. Which dot shows the location of $\frac{1}{2}$? _____

6. Which dot shows the location of $\frac{1}{3}$? _____

7. Which dot shows the location of $\frac{3}{4}$? _____

8. Choose a fraction represented on a number line above. Draw a different representation of it.

Understanding Fractions on a Number Line

Use the number lines to answer the questions.

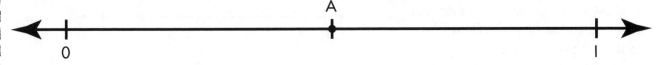

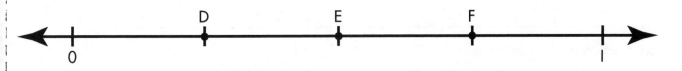

1. Label each number line with the correct fractions.

2. Which dot shows the location of $\frac{3}{6}$? _____

3. Which dot shows the location of $\frac{1}{3}$? _____

4. Which dot shows the location of $\frac{1}{6}$? _____

5. Which dot shows the location of $\frac{2}{4}$? _____

6. Which dot shows the location of $\frac{5}{6}$? _____

7. Draw a dot to represent each of the fractions below. Label them with the letter given.

   L. $\frac{2}{2}$ M. $\frac{6}{6}$ N. $\frac{3}{3}$ O. $\frac{4}{4}$

8. Choose a fraction from number 7. Explain how you decided where to place the dot on the number line.

_____

Representing Fractions on a Number Line

To show a fraction on a number line, draw a number line with 0 and 1 as the endpoints.

Then, divide it into equal parts and label them.

Finally, draw a dot on the line to represent the fraction.

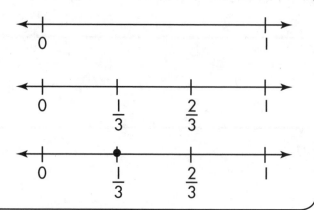

Show each fraction on the number line.

1. $\dfrac{2}{3}$ 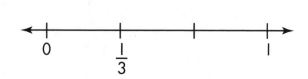	2. $\dfrac{3}{4}$ 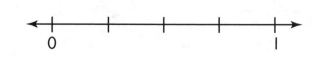
3. $\dfrac{1}{2}$ 	4. $\dfrac{5}{6}$ 
5. $\dfrac{1}{4}$	6. $\dfrac{7}{8}$

Representing Fractions on a Number Line

Remember, to show a fraction on a number line, follow these steps:

1. Draw a number line with 0 and 1 as the endpoints.

2. Divide it into equal parts and label them.

3. Draw a dot on the line to represent the fraction.

Show each fraction on the number line.

1. $\frac{1}{4}$ 0 $\frac{1}{4}$ $\frac{2}{4}$ $\frac{3}{4}$ 1	2. $\frac{3}{6}$
3. $\frac{2}{3}$ 	4. $\frac{3}{4}$ 
5. $\frac{1}{3}$ 	6. $\frac{1}{6}$
7. $\frac{2}{4}$ 	8. $\frac{5}{8}$

Representing Fractions on a Number Line

Show each fraction on the number line.

1. $\dfrac{2}{6}$	2. $\dfrac{2}{4}$
3. $\dfrac{5}{6}$	4. $\dfrac{6}{8}$
5. $\dfrac{3}{3}$	6. $\dfrac{3}{6}$
7. $\dfrac{1}{8}$	8. $\dfrac{3}{4}$
9. $\dfrac{4}{4}$	10. $\dfrac{2}{3}$

Equivalent Fractions

$$\frac{1}{2} = \frac{2}{4}$$

Fractions that equal the same amount are called equivalent fractions.

$$\frac{1}{4} = \frac{2}{8}$$

It's the same amount of the whole, the pieces are just different sizes!

Write the equivalent fractions.

1.

_____ = _____

2.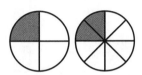

_____ = _____

3.

_____ = _____

4.

_____ = _____

5.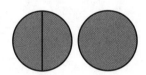

_____ = _____

6.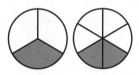

_____ = _____

7.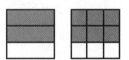

_____ = _____

8.

_____ = _____

9.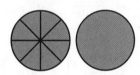

_____ = _____

10.

_____ = _____

11.

_____ = _____

12.

_____ = _____

Equivalent Fractions

Equivalent fractions name the same amount of a whole.

 $\dfrac{1}{2}$ is equivalent to $\dfrac{3}{6}$

Write each fraction. Draw a line to connect the equivalent fractions.

1.

2.

3.

4.

5.

6.

Find each equivalent fraction.

7. $\dfrac{1}{2} = \dfrac{\boxed{}}{4}$

8. $\dfrac{1}{3} = \dfrac{\boxed{}}{6}$

9. $\dfrac{3}{4} = \dfrac{\boxed{}}{8}$

10. $\dfrac{1}{2} = \dfrac{3}{\boxed{}}$

11. $\dfrac{2}{3} = \dfrac{4}{\boxed{}}$

12. $\dfrac{4}{8} = \dfrac{\boxed{}}{4}$

13. $\dfrac{2}{2} = \dfrac{3}{\boxed{}}$

14. $\dfrac{4}{6} = \dfrac{2}{\boxed{}}$

15. $\dfrac{4}{4} = \dfrac{2}{\boxed{}}$

16. $\dfrac{1}{3} = \dfrac{\boxed{}}{6}$

17. Choose two equivalent fractions from problems 7–11. Draw a picture to show how they are equivalent.

18. Choose two equivalent fractions from problems 12–16. Use number lines to show how they are equivalent.

Equivalent Fractions

Decide if each pair of fractions is equivalent. Draw an X over the pairs that are not equivalent. Draw a picture to explain why or why not.

1. $\frac{1}{4}$ and $\frac{3}{8}$

2. $\frac{2}{3}$ and $\frac{2}{4}$

3. $\frac{4}{8}$ and $\frac{1}{2}$

4. $\frac{2}{2}$ and $\frac{6}{6}$

5. $\frac{1}{2}$ and $\frac{3}{6}$

6. $\frac{6}{8}$ and $\frac{4}{6}$

7. $\frac{3}{4}$ and $\frac{2}{3}$

8. $\frac{3}{3}$ and $\frac{4}{4}$

9. $\frac{2}{3}$ and $\frac{3}{6}$

Comparing Fractions

When comparing fractions, look at the numerators and denominators.

If the **numerators** are the same, compare the denominators. The fraction with the smaller denominator is divided into fewer, larger pieces. So, it is the greater fraction.

$\frac{1}{8}$ $\frac{1}{2}$

If the **denominators** are the same, compare the numerators. The fraction with the bigger numerator has more of the same-size pieces. So, it is the greater fraction.

$\frac{1}{4}$ $\frac{3}{4}$

Identify each fraction. Circle the greater fraction.

1.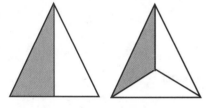

☐/☐   ☐/☐

2.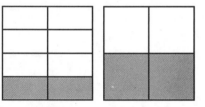

☐/☐   ☐/☐

3.

☐/☐   ☐/☐

4.

☐/☐   ☐/☐

5.

☐/☐   ☐/☐

6.

☐/☐   ☐/☐

7.

☐/☐   ☐/☐

8.

☐/☐   ☐/☐

Comparing Fractions

Circle the comparisons in each set that are not true. Rewrite the false comparisons so that they are true. Draw pictures to prove your corrections.

1. $\frac{1}{4} < \frac{1}{3}$ $\frac{1}{6} > \frac{1}{9}$ $\frac{1}{5} > \frac{1}{2}$ $\frac{1}{3} < \frac{1}{5}$ $\frac{1}{7} > \frac{1}{10}$

2. $\frac{2}{5} > \frac{4}{5}$ $\frac{6}{7} > \frac{2}{7}$ $\frac{1}{3} < \frac{2}{3}$ $\frac{4}{8} > \frac{6}{8}$ $\frac{1}{9} < \frac{4}{9}$

3. $\frac{1}{2} < \frac{1}{4}$ $\frac{3}{4} > \frac{1}{4}$ $\frac{5}{6} < \frac{2}{6}$ $\frac{1}{5} > \frac{1}{10}$ $\frac{4}{5} > \frac{2}{5}$

4. $\frac{5}{8} > \frac{2}{8}$ $\frac{1}{6} < \frac{1}{10}$ $\frac{2}{9} > \frac{4}{9}$ $\frac{1}{7} > \frac{1}{12}$ $\frac{1}{2} > \frac{1}{11}$

5. $\frac{4}{5} > \frac{3}{5}$ $\frac{2}{6} < \frac{5}{6}$ $\frac{1}{3} < \frac{1}{6}$ $\frac{1}{4} > \frac{2}{4}$ $\frac{1}{5} > \frac{1}{8}$

6. $\frac{1}{7} > \frac{1}{8}$ $\frac{1}{8} > \frac{1}{4}$ $\frac{7}{9} > \frac{3}{9}$ $\frac{4}{10} < \frac{7}{10}$ $\frac{5}{6} < \frac{3}{6}$

Name _____

Comparing Fractions

Compare using >, <, or =.

1. $\frac{5}{10}$ ◯ $\frac{2}{10}$ 2. $\frac{1}{3}$ ◯ $\frac{2}{3}$ 3. $\frac{5}{8}$ ◯ $\frac{6}{8}$

4. $\frac{3}{10}$ ◯ $\frac{8}{10}$ 5. $\frac{1}{4}$ ◯ $\frac{3}{4}$ 6. $\frac{6}{7}$ ◯ $\frac{3}{7}$

7. $\frac{4}{6}$ ◯ $\frac{1}{6}$ 8. $\frac{5}{9}$ ◯ $\frac{4}{9}$ 9. $\frac{6}{11}$ ◯ $\frac{9}{11}$

10. $\frac{1}{5}$ ◯ $\frac{3}{5}$ 11. $\frac{3}{4}$ ◯ $\frac{2}{4}$ 12. $\frac{2}{3}$ ◯ $\frac{1}{3}$

13. $\frac{1}{2}$ ◯ $\frac{1}{4}$ 14. $\frac{1}{3}$ ◯ $\frac{2}{3}$ 15. $\frac{3}{4}$ ◯ $\frac{1}{4}$

16. $\frac{1}{4}$ ◯ $\frac{1}{3}$ 17. $\frac{6}{8}$ ◯ $\frac{2}{8}$ 18. $\frac{2}{3}$ ◯ $\frac{2}{9}$

19. $\frac{2}{6}$ ◯ $\frac{2}{3}$ 20. $\frac{1}{5}$ ◯ $\frac{2}{5}$ 21. $\frac{3}{4}$ ◯ $\frac{3}{8}$

22. Choose 3 problems from above. Draw a picture under each problem to explain the comparison.

Elapsed Time

To calculate elapsed time, or time that has passed, try these methods:

The school play started at 3:00. It lasted 2 hours 10 minutes. What time did it end?

Step 1: Start at 3:00. Add or count the hours.

Step 2: Add or count the minutes.

It will end at 5:10.

Step 1: Draw a number line that marks the starting and ending times. Write the start and end times if you know them.

Step 2: Draw a line to show the amount of time to get from the start time to the next hour, and to get to the previous hour from the end time. Write how much time each space covers.

Step 3: Mark how many hours are represented in the middle.

Step 4: Add the times together.

Use the clocks to help you find the elapsed time.

1. Harry rode his bike from 4:00 to 4:58. How long did he ride his bike?

2. Jeff's game started at 2:00. It ended at 3:43. How long did the game last?

3. Zoe's favorite movie starts at 7:15. It will last for 2 hours and 18 minutes. What time will the movie end?

4. Pollo's cat disappeared at 3:12. Pollo found him at 4:26. How long was his cat lost?

5. The train leaves for New York at 8:05. The ride is 3 hours 36 minutes long. What time will the train arrive in New York?

6. Dawn started reading at 4:27. She read until 6:06. How long did Dawn read?

Elapsed Time

Use the clocks to help you find the elapsed time.

1. The dancing dogs came on at 6:15. They danced until 6:42. How long did they dance?

_____

2. The clowns started at 7:03. They rode bikes for 47 minutes. What time did they end?

_____

3. The lion tamer started at 4:21. He was on stage until 5:29. How long did he perform?

_____

4. The elephants came on at 5:19. They performed for 1 hour and 4 minutes. What time did they finish?

_____

5. The tightrope walker began at 8:44. She finished at 9:28. How long was she walking?

_____

6. The human rocket came on at 9:45. He shot off the stage 13 minutes later. What time did he leave?

_____

7. Gabe's family left their house at 5:27 for the show. They sat in their seats at 5:56. How long did it take them to get to the circus?

_____

8. The show started at 6:02. It finally ended at 9:59. How long did the entire show take?

_____

Elapsed Time

Read each problem. Use clocks or number lines to solve.

1. It is 11:54. The class will return from lunch at 12:25. How many minutes until the class returns?

    _____

2. It is 7:11. The movie starts at 8:30. How long until the movie starts?

    _____

3. It is 4:43. Dinner is at 5:15. How long until dinner?

    _____

4. The pizza takes 35 minutes to bake. I put it in the oven at 5:18. When will it be done?

    _____

5. The soccer field is 26 minutes away. If we leave at 3:53, when do we arrive at the field?

    _____

6. Jack's new game shuts down after 45 minutes. He started the game at 4:48. When will the game turn itself off?

    _____

7. Room 12 goes to music at 2:35. Music is done at 3:15. How long is music?

    _____

8. The video is 49 minutes long. We start watching it at 7:22. What time will the video be done?

    _____

9. It is 8:07 The bus comes at 8:30. How long until the bus comes?

    _____

10. Sheila ran 2 miles in 16 minutes. She started at 2:49. When did she finish?

    _____

Mass and Liquid Volume

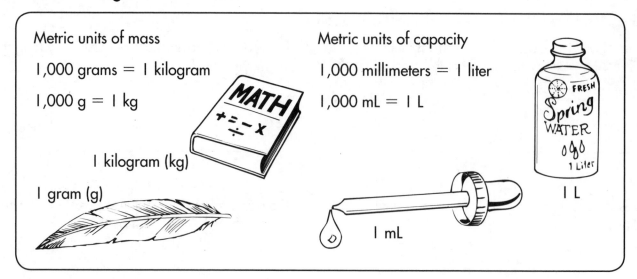

Metric units of mass

1,000 grams = 1 kilogram

1,000 g = 1 kg

1 kilogram (kg)

1 gram (g)

Metric units of capacity

1,000 millimeters = 1 liter

1,000 mL = 1 L

1 mL

1 L

Circle the correct unit to measure the following items.

1.

g   kg

2.

mL   L

3.

g   kg

4.

g   kg

5.

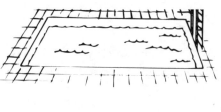

mL   L

6.

mL   L

7.

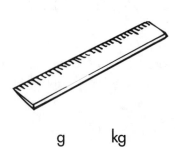

g   kg

Solve.

8. Mrs. Murphy filled a bucket to mop the floor. Does her bucket probably hold 10 milliliters or 10 liters of water?

9. A party hat has a mass of 30 grams. What is the mass of a set of 8 party hats?

10. Tony has a small pool that holds 150 liters of water. He has filled it with 87 liters of water so far. How many more liters can he add to the pool?

Mass and Liquid Volume

Circle the best estimate.

1.

    g   kg

2.

    mL   L

3.

    g   kg

4.

    g   kg

5.

    mL   L

6.

    g   kg

7.

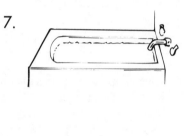

    mL   L

Solve.

8. Jenna packed 15 kilograms of apples equally into 3 bags. How many kilograms of apples were in each bag?

9. An adult weighs about 63 kilograms. An adult male moose weighs 396 kilograms. How much more does a moose weigh than a human?

10. Pablo's dog's water dish had 250 milliliters of water. His dog drank 104 milliliters; then, his dad added 58 milliliters. How many milliliters of water does the dog's dish have now?

Mass and Liquid Volume

Choose the best unit to measure each item (**g**, **kg**, **mL**, or **L**).

1. a bathtub _____

2. a medicine dropper _____

3. a bird _____

4. a bowl of soup _____

5. a baseball _____

6. a large bottle of juice _____

7. a golf cart _____

8. a cupcake _____

9. a hat _____

10. a pen _____

Solve.

11. Mr. Diaz bought drinks for a party. Did he buy 14 liters or 14 milliliters of drinks? How do you know?

12. Taylor, Quan, and India each have a book bag that weighs 4 kilograms. How many kilograms do their book bags weigh in all?

13. Hunter has a collection of bouncy balls. Each ball weighs 9 grams. His whole collection weighs 450 grams. How many bouncy balls are in Hunter's collection?

14. The grocery store stocked 2-liter bottles on the shelves. Each shelf can hold 8 bottles. There are 5 shelves in the section. How many liters can the store display at one time?

15. Lily dished up 234 milliliters of soup. Her brother served himself 281 milliliters of soup. If the pot started with 600 milliliters of soup, is there enough left for Lily's mom to have 250 milliliters? Why or why not?

Bar Graphs and Pictographs

Pictographs use pictures to compare information.

Each ★ stands for 2 awards.

Good Deed Awards

Mark	★ ★ ★ ★
Gwen	★ ★ ★ ★ ★
Emily	★ ★ ★
Katie	★
Billy	★ ★ ★

Bar graphs use bars to compare information.

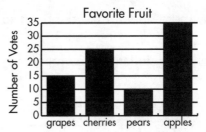

Use the information in the charts to complete the graphs. Then, use the graphs to answer the questions.

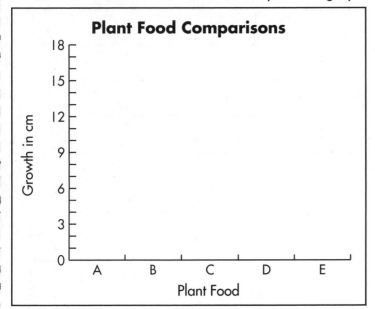

Plant Food Comparisons

Plant Food	Growth (in cm)
A	6
B	13
C	17
D	6
E	9

Flowers We Planted

Rm 102	
Rm 103	
Rm 104	
Rm 105	
✿ = 5 flowers	

Classroom	Flowers Planted
Room 102	35
Room 103	25
Room 104	20
Room 105	25

Bar Graphs and Pictographs

Use the information in the charts to complete the graphs. Then, use the graphs to answer the questions.

Animal	Votes
elephant	12
sea lion	15
bird	6
snake	18

1. How many more students voted for sea lions than birds? _____

2. How many more students voted for snakes than elephants? _____

3. How many votes did elephants and birds receive altogether? _____

4. How many votes did sea lions and snakes receive altogether? _____

5. How many votes were recorded altogether? _____

Favorite Animals

elephant	
sea lion	
bird	
snake	

Each 😊 = 3 votes.

Food Item	Number Sold
hot dogs	30
hamburgers	50
chips	50
fries	30
fruit bowls	60
ice cream	40

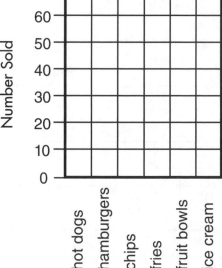

Concession Stand Sales

6. How many fruit bowls and ice-cream treats were sold altogether? _____

7. Which two items sold the least? _____

8. Which item was sold the most? _____

9. How many more hamburgers were sold than hot dogs? _____

10. How many more chips were sold than fries? _____

Bar Graphs and Pictographs

Use the information in the charts to complete the graphs. Then, use the graphs to answer each question.

Grade Level	Pounds Donated
K	30
1st	40
2nd	40
3rd	60
4th	70
5th	20

Month	Number of Students
February	30
April	50
June	70
September	10

Lindy Elementary Food Drive

Each 🗍 stands for 10 lb. of donated food.

Rocket Day Fun!

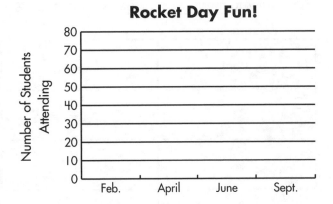

1. Which grade level had the most donations? _____

2. Which grade level had only 20 pounds of food donated? _____

3. What was the total amount of food donated for the entire school? _____

4. How much more did fourth grade donate than fifth grade? _____

5. Who donated more, first or third grade? _____

6. How many more pounds did kindergarten donate than fifth grade? _____

7. Which month had the greatest attendance at Rocket Day? _____

8. How many more students attended Rocket Day in February than in September? _____

9. Which month had 50 students attend? _____

10. How many more students attended in June than in April? _____

11. What was the total number of students that attended Rocket Day in all? _____

12. Which month had the least number of students attending? _____

Line Plots

A **line plot** is a type of graph that shows information on a number line.

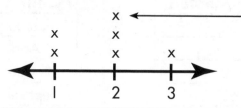

Line plots are useful for showing frequency, or the number of times something is repeated.

1. Use a ruler to measure 8 things to the nearest $\frac{1}{2}$ inch. Record your data on the table.

Item	Length	Item	Length

2. Use the data from the table to make a line plot.

- First, look at the data and decide what numbers you will need to include.

- Then, mark each number on the line plot and label it. Do not leave out numbers in between, even if they have no data!

- Finally, mark an X on the line plot to represent each piece of data.

Line Plots

1. Use a ruler to measure 10 things to the nearest $\frac{1}{2}$ inch. Record your data on the table.

Item	Length	Item	Length

2. Use the data from the table to make a line plot. Remember to look at your data to see what numbers you need to represent. Then, divide and label the line. Mark each data point with an X.

Line Plots

1. Use a ruler to measure 10 things to the nearest $\frac{1}{4}$ inch. Record your data on the table.

Item	Length	Item	Length

2. Use the data from the table to make a line plot.

Understanding Area

Area is the number of square units it takes to cover the surface of a figure.

To find area, count the number of squares it takes to cover the shape. The squares must touch along the edges with no overlap and no gaps.

Area is measured in *square units*, such as *square inches* or *square centimeters*. If the unit is not known, you can use square units as shown in the example.

Area (A) = 6 square units

Find the area of each figure.

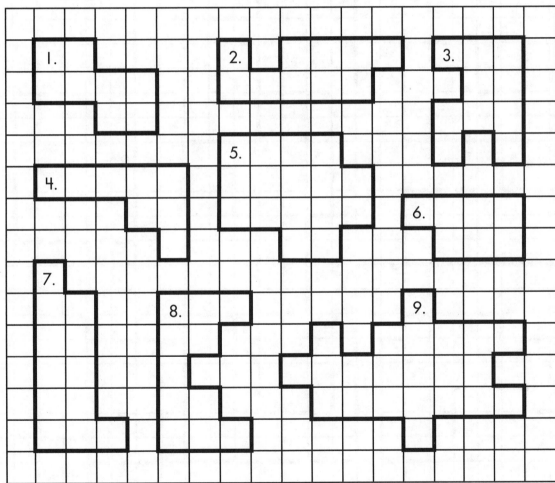

1. A = _____ sq. units

2. A = _____ sq. units

3. A = _____ sq. units

4. A = _____ sq. units

5. A = _____ sq. units

6. A = _____ sq. units

7. A = _____ sq. units

8. A = _____ sq. units

9. A = _____ sq. units

Understanding Area

Remember, **area** is the number of square units with no gaps or overlaps in a figure.

When measuring area, it helps to mark each square as you count it.

Find the area of each figure.

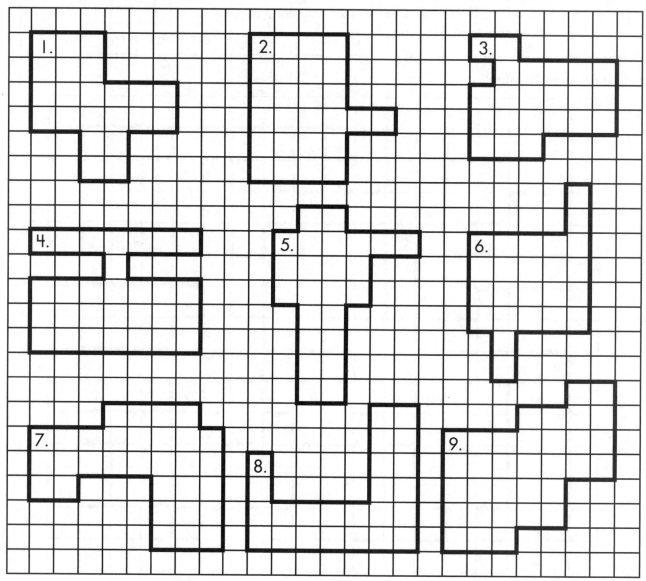

1. A = _____ sq. units

2. A = _____ sq. units

3. A = _____ sq. units

4. A = _____ sq. units

5. A = _____ sq. units

6. A = _____ sq. units

7. A = _____ sq. units

8. A = _____ sq. units

9. A = _____ sq. units

Understanding Area

Find the area of each figure.

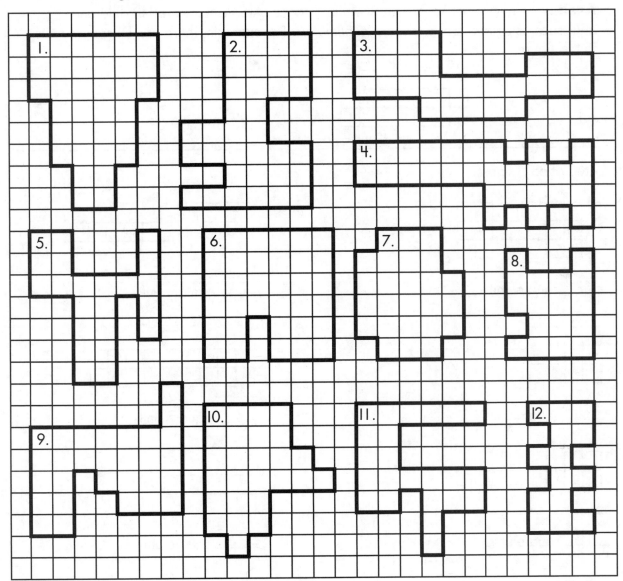

1. A = _____ sq. units

2. A = _____ sq. units

3. A = _____ sq. units

4. A = _____ sq. units

5. A = _____ sq. units

6. A = _____ sq. units

7. A = _____ sq. units

8. A = _____ sq. units

9. A = _____ sq. units

10. A = _____ sq. units

11. A = _____ sq. units

12. A = _____ sq. units

Finding Area

Area is the number of square units within a space. Area is measured in different units such as *square feet* or *square centimeters*. For example, there are 14 square units in this figure.

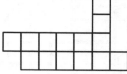

Area (A) = 14 square units

Find the area of each figure.

1.

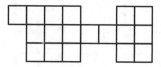

_____ square units

2.

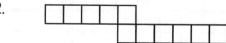

_____ square units

3.

_____ square units

4.

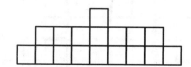

_____ square units

5.

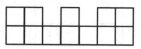

_____ square units

6.

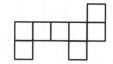

_____ square units

7.

_____ square units

8.

_____ square units

9.

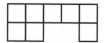

_____ square units

10.

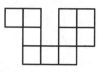

_____ square units

Finding Area

The area of a figure can be determined by counting its number of square units.

Area (A) = 9 square units

Find the area of each figure.

1.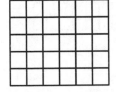

_____

2.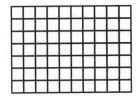

_____

3.

_____

4.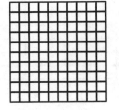

_____

5.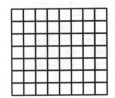

_____

6.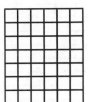

_____

7.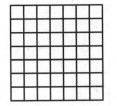

_____

8.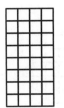

_____

9.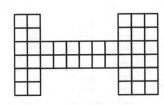

_____

10.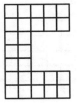

_____

11.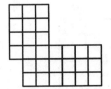

_____

12.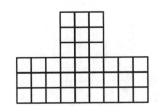

_____

Finding Area

Find the area of each figure.

1.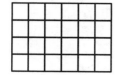

_____

2.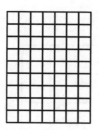

_____

3.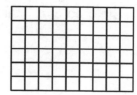

_____

4.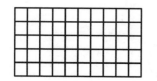

_____

5.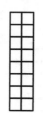

_____

6.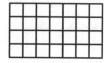

_____

7.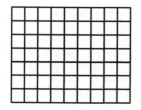

_____

8.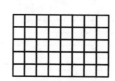

_____

9.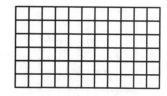

_____

10.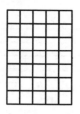

_____

11.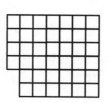

_____

12.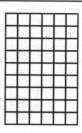

_____

13.

_____

14.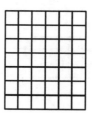

_____

15.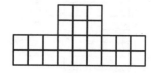

_____

Finding Area by Tiling

You can find area by *tiling* a shape, or dividing it into squares.

To tile a rectangular shape, first draw lines from top to bottom. Then, draw lines from left to right.

Finally, count the squares to find the area.

Tile each rectangle. Then, find the area. The first one is done for you.

1.

A = **9 sq. units**

2.

A = _____

3.

A = _____

4.

A = _____

5.

A = _____

6.

A = _____

7.

A = _____

8.

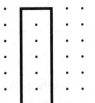

A = _____

9.

A = _____

10.

A = _____

Finding Area by Tiling

Tile each rectangle. Then, find the area.

1.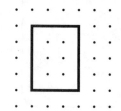

A = _____

2.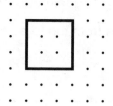

A = _____

3.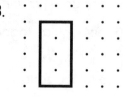

A = _____

Tiling breaks a shape up into an array of squares.

Just like with arrays, you can multiply the sides to find the total quickly.

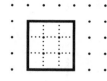

$3 \times 3 = 9$

A = 9 sq. units

Find the area using tiling and multiplication.

4.

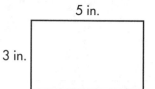

5 in.

3 in.

A = _____

5.

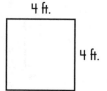

4 ft.

4 ft.

A = _____

6.

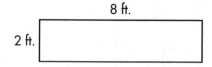

8 ft.

2 ft.

A = _____

7.

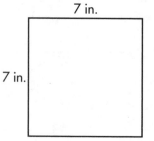

7 in.

7 in.

A = _____

8.

9 in.

5 in.

A = _____

9.

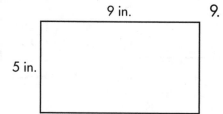

3 yd.

10 yd.

A = _____

Finding Area by Tiling

Find the area of each rectangle.

1.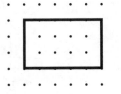

A = _____

2.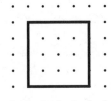

A = _____

3.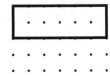

A = _____

4.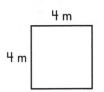

A = _____

5.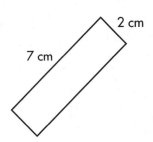

A = _____

6.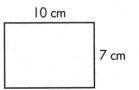

A = _____

7.
4 m

4 m

A = _____

8.
2 cm

7 cm

A = _____

9.
30 mm

2 mm

A = _____

10.
10 cm

7 cm

A = _____

11.
4 mm

20 mm

A = _____

12.
5 mm   5 mm

A = _____

Finding Area by Multiplying

The **area** is the amount of square units within a shape.

You can find area by tiling and counting the squares.

Or, you can multiply the length and width of a rectangle to find the area.

2 cm × 8 cm = 16 square centimeters

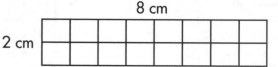

Find the area of each rectangle by tiling and by multiplying.

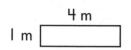

4 m

1 m

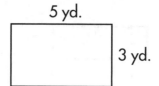

5 yd.

3 yd.

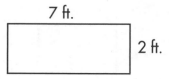

7 ft.

2 ft.

1. A = _____ × _____

   A = _____ sq. m

2. A = _____ × _____

   A = _____ sq. yd.

3. A = _____ × _____

   A = _____ sq. ft.

Find the area of each rectangle by multiplying.

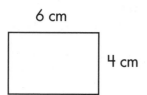

6 cm

4 cm

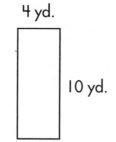

4 yd.

10 yd.

10 ft.

1 ft.

4. A = _____ × _____

   A = _____ sq. cm

5. A = _____ × _____

   A = _____ sq. yd.

6. A = _____ × _____

   A = _____ sq. ft.

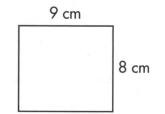

10 m

5 m

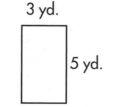

3 yd.

5 yd.

9 cm

8 cm

7. A = _____ × _____

   A = _____ sq. m

8. A = _____ × _____

   A = _____ sq. yd.

9. A = _____ × _____

   A = _____ sq. cm

Finding Area by Multiplying

> Remember, you can find the area of a rectangle by tiling, or multiplying the length and width.

Find the area of each rectangle by tiling and by multiplying.

1.
9 ft.
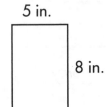
9 ft.

A = _____ × _____

A = _____

2.
5 in.
8 in.

A = _____ × _____

A = _____

3.
6 in.
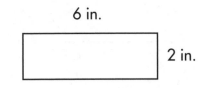
2 in.

A = _____ × _____

A = _____

Find the area of each rectangle by multiplying.

4.
8 cm
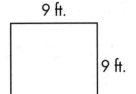
2 cm

A = _____ × _____

A = _____

5.
6 yd.
6 yd.

A = _____ × _____

A = _____

6.
9 ft.
4 ft.

A = _____ × _____

A = _____

7.
5 in.
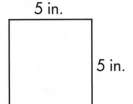
5 in.

A = _____ × _____

A = _____

8.
5 cm
4 cm
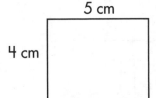

A = _____ × _____

A = _____

9.
3 mm
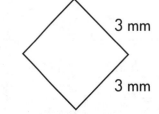
3 mm

A = _____ × _____

A = _____

Finding Area by Multiplying

Find the area of each figure.

1.
6 cm
4 cm

A = _____

2.
2 in.
2 in.

A = _____

3.
3 ft.
10 ft.

A = _____

4.
8 yd.
2 yd.

A = _____

5.
5 mm
9 mm

A = _____

6.
8 m
8 m

A = _____

7.
4 in.
4 in.

A = _____

8.
10 ft.
6 ft.

A = _____

9.
3 cm
7 cm

A = _____

10.
5 m
5 m

A = _____

11.
6 yd.
2 yd.

A = _____

12.
4 yd.
3 yd.

A = _____

13.
5 in.
8 in.

A = _____

14.
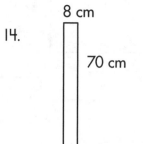
8 cm
70 cm

A = _____

15.
9 m
11 m

A = _____

Finding Area of Rectilinear Figures

> To find the area of a complex rectangular figure, divide the figure into two or more rectangles.
>
> Then, find the area of each rectangle.
>
> Add the areas together to find the total area of the shape.

Find the area of each figure.

1.

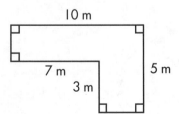

A = _____

2.

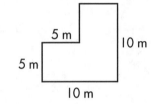

A = _____

3.

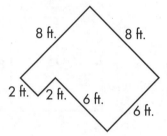

A = _____

4.

A = _____

5.

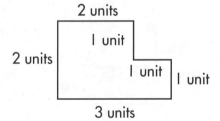

A = _____

6.

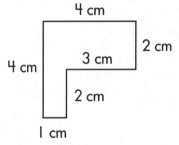

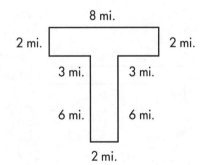

A = _____

3.MD.C.7d

Finding Area of Rectilinear Figures

Remember, to find the area of a complex rectangular figure, follow these steps:

1. Divide the figure into two or more rectangles.

2. Find the area of each rectangle.

3. Add the areas together to find the total area of the shape.

Find the area of each figure.

1.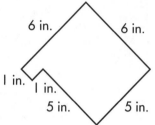

A = _____

2.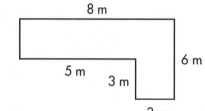

A = _____

3.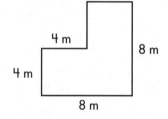

A = _____

4.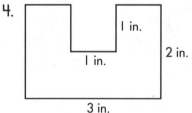

A = _____

5.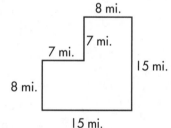

A = _____

6.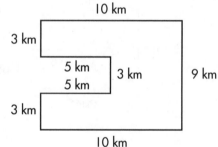

A = _____

7.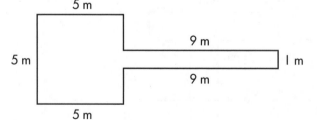

A = _____

8.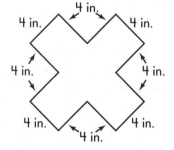

A = _____

Finding Area of Rectilinear Figures

Find the area of each figure. All measurements are in centimeters.

1.
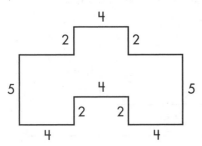

A = _____

2.
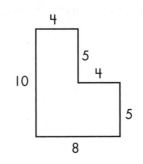

A = _____

3.
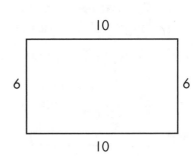

A = _____

4.
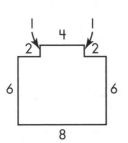

A = _____

5.
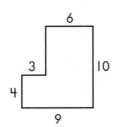

A = _____

6.
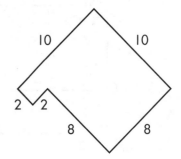

A = _____

7.
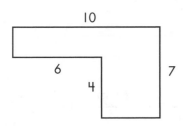

A = _____

8.
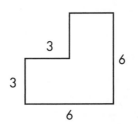

A = _____

9.
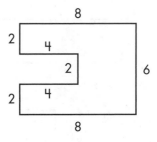

A = _____

10.
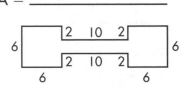

A = _____

11.
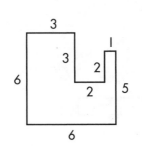

A = _____

12.
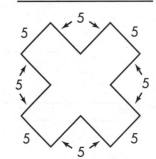

A = _____

Finding Perimeter

Perimeter is the total distance around a given figure. To find the perimeter, add the lengths of the sides of the figure.

Example: P = perimeter

P = 4 cm + 8 cm + 4 cm + 8 cm

P = 24 cm

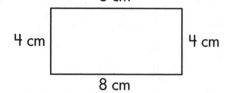

Find the perimeter of each figure.

1.

4 yd.

2 yd.

P = _____

2.

6 ft.

3 ft.

P = _____

3.

19 ft.

5 ft.

P = _____

4.

6 in.

7 in.

P = _____

5.

8 in.

8 in.

P = _____

6.

12 mm

9 mm

P = _____

7.

1 ft.

14 ft.

P = _____

8.

9 mm

3 mm

P = _____

Finding Perimeter

Find the perimeter of each figure.

1.
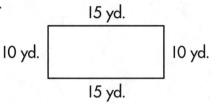

P = 15 + 10 + 15 + 10

P = _____

2.
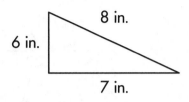

P = _____

3.
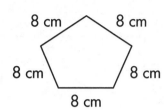

P = _____

4.
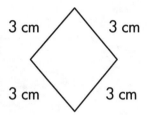

P = _____

5.
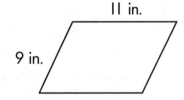

P = _____

6.
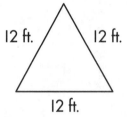

P = _____

7.
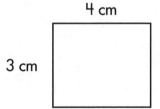

P = _____

8.
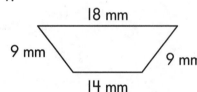

P = _____

9.

P = _____

10. Mrs. Young's tennis court is a rectangle that is 30 feet wide by 70 feet long. What is the perimeter?

P = _____

11. A square tile has sides that are 15 inches each. What is the perimeter?

P = _____

12. A hexagon has sides that are 5 millimeters. What is the perimeter?

P = _____

3.MD.D.8

Finding Perimeter

Find the perimeter of each figure.

1.

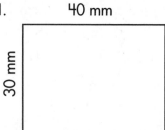

40 mm

30 mm

P = _____

2.

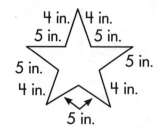

4 in. 4 in.
5 in. 5 in.
5 in.
5 in.
4 in. 4 in.
5 in.

P = _____

3.

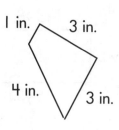

1 in. 3 in.
4 in.
3 in.

P = _____

4.

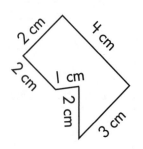

2 cm
2 cm
1 cm
4 cm
2 cm
3 cm

P = _____

5.

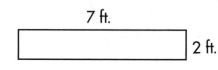

7 ft.
2 ft.

P = _____

6.

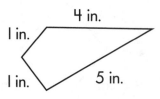

4 in.
1 in.
1 in.
5 in.

P = _____

7.

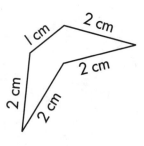

1 cm
2 cm
2 cm
2 cm
2 cm

P = _____

8.

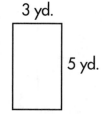

3 yd.
5 yd.

P = _____

9.

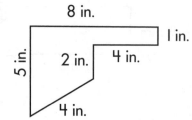

8 in.
1 in.
5 in.
2 in. 4 in.
4 in.

P = _____

10. A playing card has a length of 10 centimeters and a width of 5 centimeters. What is its perimeter?

P = _____

11. If one side of a stop sign measures 12 inches, then what is its perimeter?

P = _____

12. A flower bed is rectangular. The perimeter is 24 feet. It is 8 feet long. How wide is the flower bed?

P = _____

Classifying Polygons

Polygons are named for the number of sides they have.

    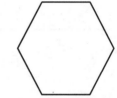

| Triangle | Quadrilateral | Pentagon | Hexagon | Octagon |
| ___ sides | ___ sides | ___ sides | ___ sides | ___ sides |

Complete the table.

Type of Polygon	Number of Sides
1.	3
2. quadrilateral	
3.	5
4. hexagon	
5. heptagon	
6.	8
7. nonagon	
8. decagon	

A.

B.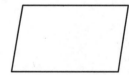

C.

Match figures A–E to the following definitions.

9. _____ a polygon with 5 sides

10. _____ a polygon with 8 sides

11. _____ a quadrilateral with opposite sides parallel

12. _____ a polygon with 3 angles

13. _____ a polygon with 6 sides

D.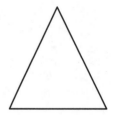

E.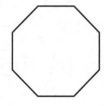

Classifying Polygons

Sort the following plane figures by drawing them in the Venn diagram below. Then, answer the questions.

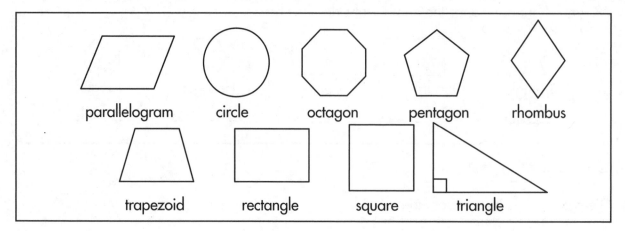

parallelogram   circle   octagon   pentagon   rhombus

trapezoid   rectangle   square   triangle

1.

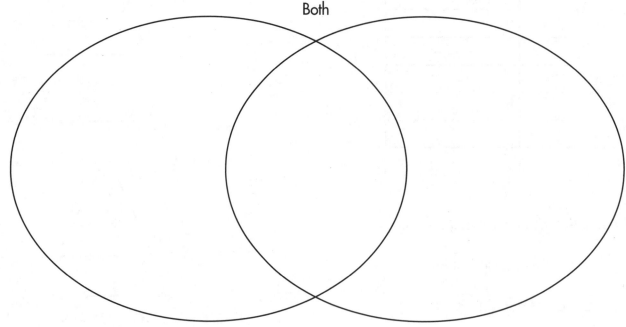

Four Sides   Equal and Opposite Sides

Both

2. Which figures do not fit into the Venn diagram? _____

   Why not? _____

3. Which figures are quadrilaterals? _____

   _____

4. Why? _____

Classifying Polygons

Complete each statement with *All, Some, No,* or *None.*

1. _____ rectangles have 4 vertices. _____ rectangles are parallelograms.
   _____ rectangles are circles.

2. _____ quadrilaterals have 4 sides. _____ quadrilaterals are rectangles.
   _____ quadrilaterals have just 3 vertices.

3. _____ polygons are quadrilaterals. _____ polygons have equal sides.
   _____ polygons have curved sides.

4. _____ pentagons have 5 sides. _____ hexagons have more than 5 sides.
   _____ octagons have less than 5 sides.

5. _____ triangles have 4 vertices. _____ triangles are quadrilaterals.
   _____ triangles have 3 vertices.

Look at the diagrams to complete the statements using *All, Some, No,* or *None.*

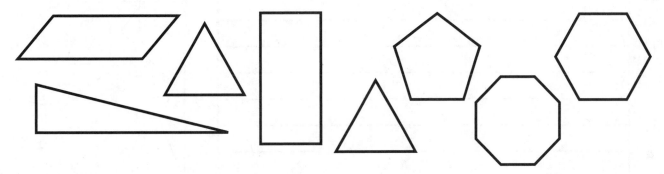

6. _____ of the polygons in this diagram have less than 3 vertices.

7. _____ of the polygons are quadrilaterals.

8. _____ polygons have three or more sides.

Recognizing Polygons

Look at the shapes on the right. Follow the directions.

1. Color the triangles green.

2. Color the quadrilaterals red.

3. Color the pentagons blue.

4. Color the hexagons orange.

5. Color the octagons purple.

6. Count the shapes to complete the chart.

Shape	Number Found
triangle	
quadrilateral	
pentagon	
hexagon	
octagon	

7. Color the graph to show how many of each shape there are.

Number of Shapes

| 20 |
| 18 |
| 16 |
| 14 |
| 12 |
| 10 |
| 8 |
| 6 |
| 4 |
| 2 |
| 0 |

Less Than 4 Sides   Quadrilateral   More Than 4 Sides

Recognizing Polygons

Identify each type of polygon as a triangle, quadrilateral, or pentagon.

     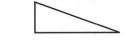

1. _____   2. _____   3. _____   4. _____

5. _____   6. _____   7. _____   8. _____

   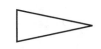   

9. _____   10. _____   11. _____   12. _____

A **parallelogram** is a special type of quadrilateral that has opposite sides that are parallel and the same length.  parallel

A **rectangle** is a parallelogram that has four right angles.   A **square** is a rectangle with four sides equal in length.

Identify each type of polygon as a parallelogram, rectangle, or square.

13. _____

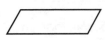

14. _____

15. _____

16. _____

Recognizing Polygons

A **quadrilateral** is a closed figure with four sides and four angles. Make four different quadrilaterals. Record your figures here.

1.

2.

3.

4.

What makes each of these figures a quadrilateral? _____

There are special types of quadrilaterals.

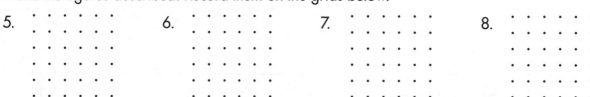

- A **trapezoid** is a quadrilateral with just one set of parallel sides.
- A **parallelogram** is a quadrilateral with two sets of parallel sides.
- A **rectangle** is a parallelogram with four right angles.
- A **square** is a rectangle with four sides of equal length.

Make the figures described. Record them on the grids below.

5.

a trapezoid

6.

a parallelogram

7.

a rectangle

8.

a square

9.

a parallelogram
that isn't a rectangle

10.

a parallelogram
that is a square

11.

a quadrilateral that
is a trapezoid

12.

a rectangle that
is a square

13.

a rectangle that
isn't a square

14.

a quadrilateral that
isn't a trapezoid or
parallelogram

15.

a parallelogram
that is a rectangle

16.

a square that
is a rectangle

Partitioning Shapes

Shapes can be **partitioned** into parts with equal areas.

Each area is a fraction of the whole.

  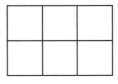

Divide each shape into 4 equal parts. Label each area with a fraction.

1.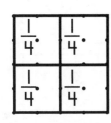

2.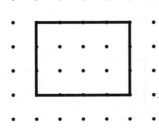

3.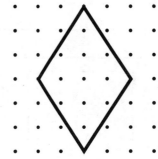

Divide each shape into 6 equal parts. Label each area with a fraction.

4.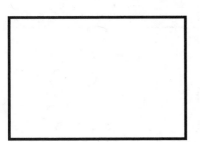

5.

6.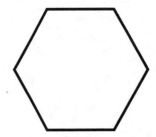

Name _____

Partitioning Shapes

Tell how each shape is partitioned.

1. 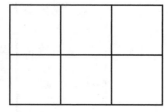   _____

2. 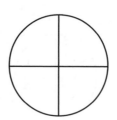   _____

3. 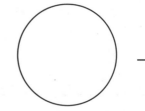   _____

Partition each shape into 2, 4, and 6 equal areas. Tell the fraction of each area.

	2	4	6
4.	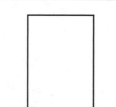 ___	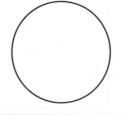 ___	 ___
5.	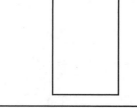 ___	___	___
6.	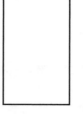 ___	___	___
7.	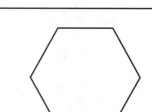 ___	___	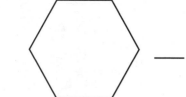 ___

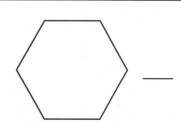

Partitioning Shapes

Draw 3 different quadrilaterals. Partition at least 1 shape into 4 equal parts.

1.

2.

3.

4. The parts are all _____.

Draw 3 different pentagons. Partition at least 1 shape into 2 equal parts.

5.

6.

7.

8. The parts are all _____.

Draw 3 different hexagons. Partition at least 1 shape into 3 equal parts.

9.

10.

11.

12. The parts are all _____.

Draw 3 different octagons. Partition at least 1 shape into 8 equal parts.

13.

14.

15.

16. The parts are all _____.

Answer Key

Name _____ 3.OA.A.1

Understanding Multiplication

To multiply means to use repeated addition. It is more easily understood if you can imagine making equal groups, and then adding all of the groups together. It looks like this:

The answer to a multiplication problem is called the **product**. The numbers being multiplied are called **factors**.

4 + 4 + 4
3 groups of 4
3 × 4 ← factors
12 ← product

Add. Then, multiply.

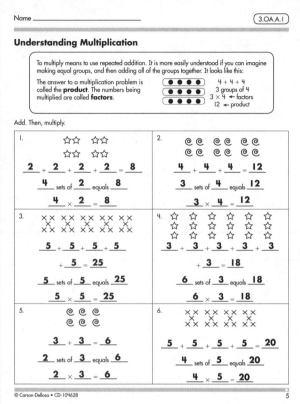

1. **2** + **2** + **2** + **2** = **8**
   4 sets of **2** equals **8**
   4 × **2** = **8**

2. **4** + **4** + **4** = **12**
   3 sets of **4** equals **12**
   3 × **4** = **12**

3. **5** + **5** + **5** + **5** + **5** = **25**
   5 sets of **5** equals **25**
   5 × **5** = **25**

4. **3** + **3** + **3** + **3** + **3** + **3** = **18**
   6 sets of **3** equals **18**
   6 × **3** = **18**

5. **3** + **3** = **6**
   2 sets of **3** equals **6**
   2 × **3** = **6**

6. **5** + **5** + **5** + **5** = **20**
   4 sets of **5** equals **20**
   4 × **5** = **20**

© Carson-Dellosa • CD-104628   5

Name _____ 3.OA.A.1

Understanding Multiplication

Write an addition and multiplication problem for each picture. Then, find the sum and the product.

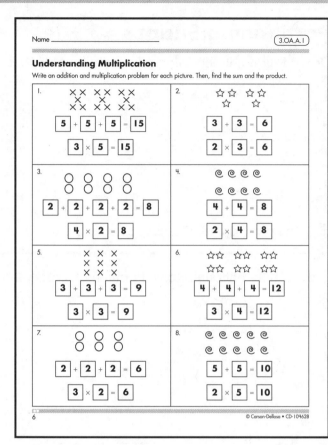

1. **5** + **5** + **5** = **15**
   3 × **5** = **15**

2. **3** + **3** = **6**
   2 × **3** = **6**

3. **2** + **2** + **2** + **2** = **8**
   4 × **2** = **8**

4. **4** + **4** = **8**
   2 × **4** = **8**

5. **3** + **3** + **3** = **9**
   3 × **3** = **9**

6. **4** + **4** + **4** = **12**
   3 × **4** = **12**

7. **2** + **2** + **2** = **6**
   3 × **2** = **6**

8. **5** + **5** = **10**
   2 × **5** = **10**

6   © Carson-Dellosa • CD-104628

Name _____ 3.OA.A.1

Understanding Multiplication

Match each multiplication problem to its addition sentence and picture. Then, solve.

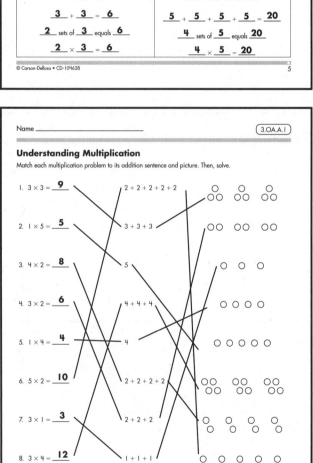

1. 3 × 3 = **9**
2. 1 × 5 = **5**
3. 4 × 2 = **8**
4. 3 × 2 = **6**
5. 1 × 4 = **4**
6. 5 × 2 = **10**
7. 3 × 1 = **3**
8. 3 × 4 = **12**

2 + 2 + 2 + 2 + 2
3 + 3 + 3
5
4 + 4 + 4
4
2 + 2 + 2 + 2
2 + 2 + 2
1 + 1 + 1

© Carson-Dellosa • CD-104628   7

Name _____ 3.OA.A.1

Multiplying Sets

To find the answer to a multiplication problem, add all of the groups together. The answer is called the **product**.

Example: 3 × 2 = ★★ ★★ ★★ = 2 + 2 + 2 = 6
(3 groups of 2)

Draw the picture. Write the multiples next to each picture. Use the picture to write an addition sentenece. Then, write the multiplication sentence.

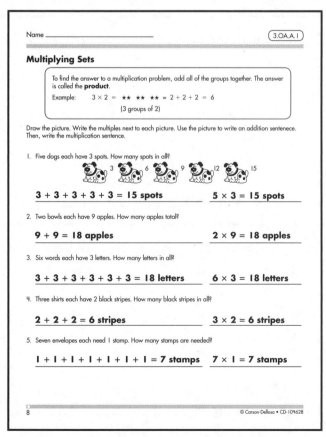

1. Five dogs each have 3 spots. How many spots in all?
   3 6 9 12 15
   3 + 3 + 3 + 3 + 3 = 15 spots   **5 × 3 = 15 spots**

2. Two bowls each have 9 apples. How many apples total?
   9 + 9 = 18 apples   **2 × 9 = 18 apples**

3. Six words each have 3 letters. How many letters in all?
   3 + 3 + 3 + 3 + 3 + 3 = 18 letters   **6 × 3 = 18 letters**

4. Three shirts each have 2 black stripes. How many black stripes in all?
   2 + 2 + 2 = 6 stripes   **3 × 2 = 6 stripes**

5. Seven envelopes each need 1 stamp. How many stamps are needed?
   1 + 1 + 1 + 1 + 1 + 1 + 1 = 7 stamps   **7 × 1 = 7 stamps**

8   © Carson-Dellosa • CD-104628

Answer Key

Name _____ 3.OA.A.1

Multiplying Sets

Draw groups to match the multiplication problem. Write the addition sentence. Then find the product.

1. 4 × 2 =	☆☆ ☆☆ ☆☆ ☆☆	=	2 + 2 + 2 + 2	= **8**
2. 3 × 3 =	☆☆☆ ☆☆☆ ☆☆☆	=	3 + 3 + 3	= **9**
3. 3 × 4 =	☆☆ ☆☆ ☆☆ ☆☆ ☆☆ ☆☆	=	4 + 4 + 4	= **12**
4. 2 × 6 =	☆☆☆☆☆☆ ☆☆☆☆☆☆	=	6 + 6	= **12**
5. 2 × 2 =	☆☆ ☆☆	=	2 + 2	= **4**
6. 2 × 5 =	☆☆☆☆☆ ☆☆☆☆☆	=	5 + 5	= **10**
7. 4 × 1 =	☆ ☆ ☆ ☆	=	1 + 1 + 1 + 1	= **4**
8. 4 × 5 =	☆☆☆☆☆ ☆☆☆☆☆ ☆☆☆☆☆ ☆☆☆☆☆	=	5 + 5 + 5 + 5	= **20**

9

Name _____ 3.OA.A.1

Multiplying Sets

Draw the sets. Then, write the multiplication problem. Solve.

Example: Five sets of three equals **15**   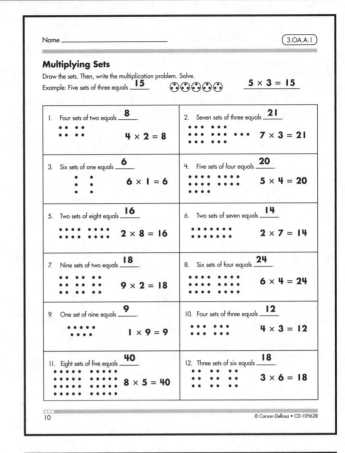   _5 × 3 = 15_

1. Four sets of two equals **8**. 4 × 2 = 8	2. Seven sets of three equals **21**. 7 × 3 = 21
3. Six sets of one equals **6**. 6 × 1 = 6	4. Five sets of four equals **20**. 5 × 4 = 20
5. Two sets of eight equals **16**. 2 × 8 = 16	6. Two sets of seven equals **14**. 2 × 7 = 14
7. Nine sets of two equals **18**. 9 × 2 = 18	8. Six sets of four equals **24**. 6 × 4 = 24
9. One set of nine equals **9**. 1 × 9 = 9	10. Four sets of three equals **12**. 4 × 3 = 12
11. Eight sets of five equals **40**. 8 × 5 = 40	12. Three sets of six equals **18**. 3 × 6 = 18

10

Name _____ 3.OA.A.2

Understanding Division

To divide means to make equal groups or to share equally. The answer to a division problem is called the **quotient**. It looks like this:

dividend   divisor   quotient

12 ÷ 3 = 4

Write each missing number.

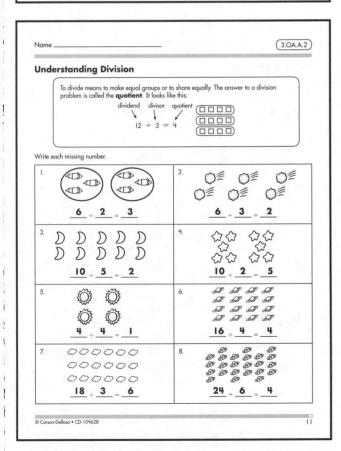

1. **6** ÷ **2** = **3**	2. **6** ÷ **3** = **2**
3. **10** ÷ **5** = **2**	4. **10** ÷ **2** = **5**
5. **4** ÷ **4** = **1**	6. **16** ÷ **4** = **4**
7. **18** ÷ **3** = **6**	8. **24** ÷ **6** = **4**

11

Name _____ 3.OA.A.2

Understanding Division

Circle to show a fair share. Write the division sentence. Write how many each person gets.

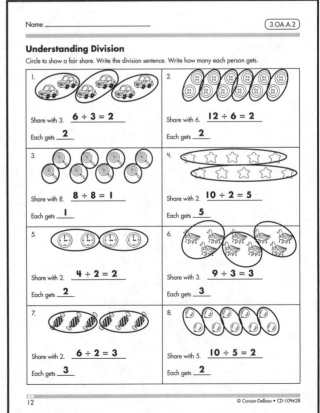

1. Share with 3. **6 ÷ 3 = 2** Each gets **2**.	2. Share with 6. **12 ÷ 6 = 2** Each gets **2**.
3. Share with 8. **8 ÷ 8 = 1** Each gets **1**.	4. Share with 2. **10 ÷ 2 = 5** Each gets **5**.
5. Share with 2. **4 ÷ 2 = 2** Each gets **2**.	6. Share with 3. **9 ÷ 3 = 3** Each gets **3**.
7. Share with 2. **6 ÷ 2 = 3** Each gets **3**.	8. Share with 5. **10 ÷ 5 = 2** Each gets **2**.

12

Answer Key

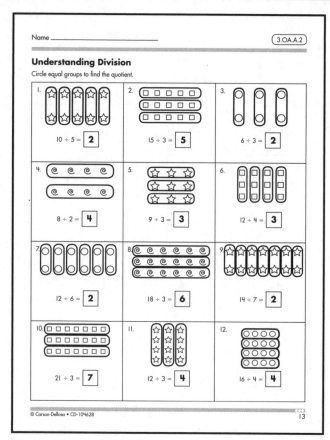

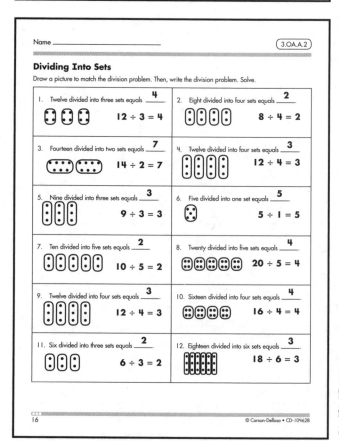

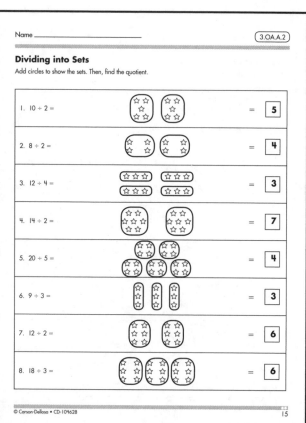

Answer Key

Answer Key

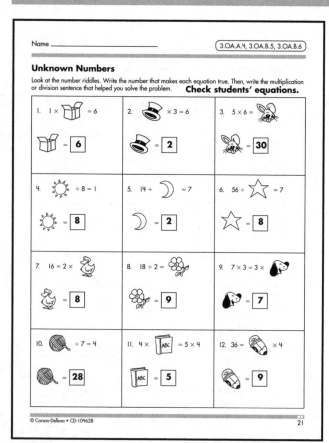

Name _____   3.OA.A.4, 3.OA.B.5, 3.OA.B.6

Unknown Numbers

Look at the number riddles. Write the number that makes each equation true. Then, write the multiplication or division sentence that helped you solve the problem. **Check students' equations.**

1. $1 \times \square = 6$
   $\square = \boxed{6}$

2. $\text{(hat)} \times 3 = 6$
   $\text{(hat)} = \boxed{2}$

3. $5 \times 6 = \text{(rabbit)}$
   $\text{(rabbit)} = \boxed{30}$

4. $\text{(sun)} \div 8 = 1$
   $\text{(sun)} = \boxed{8}$

5. $14 \div \text{(moon)} = 7$
   $\text{(moon)} = \boxed{2}$

6. $56 \div \text{(star)} = 7$
   $\text{(star)} = \boxed{8}$

7. $16 = 2 \times \text{(duck)}$
   $\text{(duck)} = \boxed{8}$

8. $18 \div 2 = \text{(flower)}$
   $\text{(flower)} = \boxed{9}$

9. $7 \times 3 = 3 \times \text{(dog)}$
   $\text{(dog)} = \boxed{7}$

10. $\text{(yarn)} \div 7 = 4$
    $\text{(yarn)} = \boxed{28}$

11. $4 \times \text{(book)} = 5 \times 4$
    $\text{(book)} = \boxed{5}$

12. $36 = \text{(pencil)} \times 4$
    $\text{(pencil)} = \boxed{9}$

© Carson-Dellosa • CD-104628   21

Name _____   3.OA.A.4, 3.OA.B.5

Unknown Numbers

Find the number that makes each set of sentences true.

1. $3 \times \underline{4} = 12, 16 \div \underline{4} = 4$
2. $16 = 4 \times \underline{4}, 8 \div \underline{4} = 2$
3. $\underline{4} \times 2 = 8, \underline{4} \div 1 = 1$
4. $10 = \underline{2} \times 5, 14 \div \underline{2} = 7$
5. $6 \times 4 = \underline{24}, \underline{24} \div 3 = 8$
6. $\underline{36} = 5 \times 9, 6 \times 6 = \underline{36}$
7. $6 \times \underline{3} = 18, 3 \times \underline{3} = 9$
8. $35 = 7 \times \underline{5}, 45 \div \underline{5} = 9$
9. $\underline{8} \times 4 = 32, 16 \div \underline{8} = 2$
10. $30 = \underline{6} \times 5, 42 \div 7 = \underline{6}$
11. $16 \div \underline{8} = 2, \underline{8} \times 7 = 56$
12. $1 = 8 \div \underline{8}, 64 \div \underline{8} = 8$
13. $20 \div 5 = \underline{4}, 36 \div \underline{4} = 9$
14. $\underline{7} = 42 \div 6, 28 \div 4 = \underline{7}$
15. $\underline{12} \div 4 = 3, 6 \times 2 = \underline{12}$
16. $2 = \underline{14} \div 7, \underline{14} = 7 \times 2$
17. $28 \div \underline{4} = 7, \underline{4} \times 5 = 20$
18. $4 = 36 \div \underline{9}, 3 \times 3 = \underline{9}$
19. $\underline{6} \div 2 = 3, 54 \div \underline{6} = 9$
20. $5 = \underline{15} \div 3, \underline{15} = 3 \times 5$
21. $\underline{7} \times 3 = 21, 28 \div \underline{7} = 4$
22. $72 = 9 \times \underline{8}, 48 \div \underline{8} = 6$
23. $\underline{40} \div 4 = 10, 8 \times 5 = \underline{40}$
24. $8 = \underline{16} \div 2, 2 \times 8 = \underline{16}$

22   © Carson-Dellosa • CD-104628

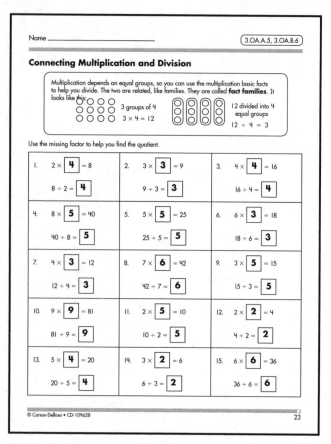

Name _____   3.OA.A.5, 3.OA.B.6

Connecting Multiplication and Division

Multiplication depends on equal groups, so you can use the multiplication basic facts to help you divide. The two are related, like families. They are called **fact families**. It looks like this:

3 groups of 4
$3 \times 4 = 12$

12 divided into 4 equal groups
$12 \div 4 = 3$

Use the missing factor to help you find the quotient.

1. $2 \times \boxed{4} = 8$
   $8 \div 2 = \boxed{4}$

2. $3 \times \boxed{3} = 9$
   $9 \div 3 = \boxed{3}$

3. $4 \times \boxed{4} = 16$
   $16 \div 4 = \boxed{4}$

4. $8 \times \boxed{5} = 40$
   $40 \div 8 = \boxed{5}$

5. $5 \times \boxed{5} = 25$
   $25 \div 5 = \boxed{5}$

6. $6 \times \boxed{3} = 18$
   $18 \div 6 = \boxed{3}$

7. $4 \times \boxed{3} = 12$
   $12 \div 4 = \boxed{3}$

8. $7 \times \boxed{6} = 42$
   $42 \div 7 = \boxed{6}$

9. $3 \times \boxed{5} = 15$
   $15 \div 3 = \boxed{5}$

10. $9 \times \boxed{9} = 81$
    $81 \div 9 = \boxed{9}$

11. $2 \times \boxed{5} = 10$
    $10 \div 2 = \boxed{5}$

12. $2 \times \boxed{2} = 4$
    $4 \div 2 = \boxed{2}$

13. $5 \times \boxed{4} = 20$
    $20 \div 5 = \boxed{4}$

14. $3 \times \boxed{2} = 6$
    $6 \div 3 = \boxed{2}$

15. $6 \times \boxed{6} = 36$
    $36 \div 6 = \boxed{6}$

© Carson-Dellosa • CD-104628   23

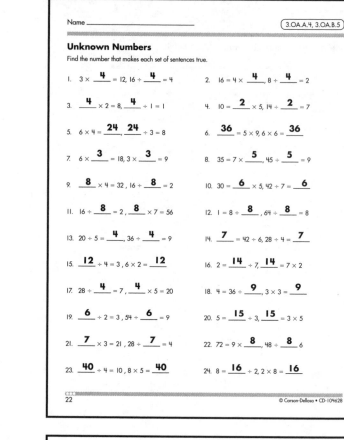

Name _____   3.OA.A.4, 3.OA.A.5, 3.OA.B.6

Connecting Multiplication and Division

Use what you know about multiplication to find the quotients. Write the related multiplication sentence.

1. $12 \div 6 = \boxed{2}$
   $6 \times 2 = 12$

2. $24 \div 4 = \boxed{6}$
   $4 \times 6 = 24$

3. $40 \div 5 = \boxed{8}$
   $5 \times 8 = 40$

4. $16 \div 4 = \boxed{4}$
   $4 \times 4 = 16$

5. $21 \div 7 = \boxed{3}$
   $7 \times 3 = 21$

6. $9 \div 3 = \boxed{3}$
   $3 \times 3 = 9$

7. $36 \div 6 = \boxed{6}$
   $6 \times 6 = 36$

8. $24 \div 8 = \boxed{3}$
   $8 \times 3 = 24$

9. $20 \div 4 = \boxed{5}$
   $4 \times 5 = 20$

10. $15 \div 5 = \boxed{3}$
    $5 \times 3 = 15$

11. $12 \div 4 = \boxed{3}$
    $4 \times 3 = 12$

12. $25 \div 5 = \boxed{5}$
    $5 \times 5 = 25$

13. $9\overline{)27}$ → 3
    $9 \times 3 = 27$

14. $9\overline{)36}$ → 4
    $9 \times 4 = 36$

15. $9\overline{)81}$ → 9
    $9 \times 9 = 81$

16. $6\overline{)54}$ → 9
    $6 \times 9 = 54$

17. $9\overline{)63}$ → 7
    $9 \times 7 = 63$

18. $5\overline{)45}$ → 9
    $5 \times 9 = 45$

19. $7\overline{)56}$ → 8
    $7 \times 8 = 56$

20. $7\overline{)49}$ → 7
    $7 \times 7 = 49$

21. $8\overline{)64}$ → 8
    $8 \times 8 = 64$

22. $7\overline{)42}$ → 6
    $7 \times 6 = 42$

24   © Carson-Dellosa • CD-104628

Answer Key

Connecting Multiplication and Division

Jamie does not have the times tables memorized yet.

Show two or more ways to solve each multiplication problem using properties of operations, such as using fact families or breaking it apart into easier problems.

1. 6 × 6 =

2. 8 × 9 =

3. 4 × 7 =

4. 7 × 6 =

5. 3 × 9 =

6. 7 × 8 =

7. 3 × 8 =

8. 4 × 9 =

9. 8 × 8 =

10. 2 × 3 × 3 =

11. 4 × 2 × 3 =

12. 5 × 4 × 2 =

Answers will vary.

Multiplication Fluency with Factors 0–5

Solve each problem.

1. 5 ×3 = **15**	2. 3 ×4 = **12**	3. 5 ×4 = **20**	4. 6 ×3 = **18**	5. 1 ×0 = **0**	6. 5 ×6 = **30**
7. 9 ×3 = **27**	8. 2 ×0 = **0**	9. 2 ×4 = **8**	10. 2 ×3 = **6**	11. 8 ×2 = **16**	12. 4 ×8 = **32**
13. 3 ×3 = **9**	14. 4 ×3 = **12**	15. 4 ×1 = **4**	16. 3 ×0 = **0**	17. 5 ×7 = **35**	18. 2 ×0 = **0**
19. 2 ×1 = **2**	20. 1 ×7 = **7**	21. 9 ×2 = **18**	22. 1 ×0 = **0**	23. 4 ×5 = **20**	24. 1 ×4 = **4**
25. 8 ×5 = **40**	26. 5 ×2 = **10**	27. 5 ×5 = **25**			

Multiplication Fluency with Factors 0–5

Solve each problem.

1. 2 ×5 = **10**	2. 5 ×8 = **40**	3. 5 ×3 = **15**	4. 8 ×4 = **32**	5. 3 ×4 = **12**	6. 7 ×2 = **14**
7. 7 ×5 = **35**	8. 1 ×4 = **4**	9. 3 ×0 = **0**	10. 2 ×2 = **4**	11. 8 ×3 = **24**	12. 4 ×3 = **12**
13. 4 ×6 = **24**	14. 5 ×2 = **10**	15. 4 ×5 = **20**	16. 2 ×9 = **18**	17. 5 ×5 = **25**	18. 5 ×6 = **30**
19. 4 ×2 = **8**	20. 4 ×9 = **36**	21. 9 ×3 = **27**	22. 4 ×4 = **16**	23. 3 ×7 = **21**	24. 8 ×2 = **16**
25. 6 ×2 = **12**	26. 3 ×6 = **18**	27. 2 ×0 = **0**	28. 4 ×7 = **28**	29. 3 ×2 = **6**	30. 9 ×5 = **45**
31. 5 ×1 = **5**	32. 2 ×3 = **6**	33. 3 ×1 = **3**			

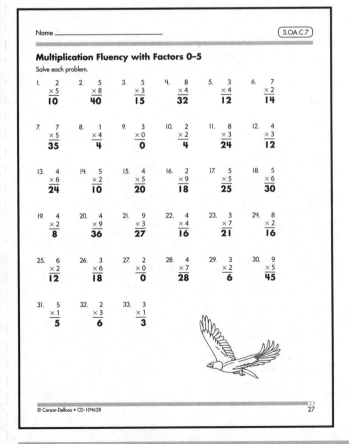

Multiplication Fluency with Factors 0–5

Solve each problem.

1. 2 ×5 = **10**	2. 5 ×8 = **40**	3. 6 ×3 = **18**	4. 8 ×4 = **32**	5. 3 ×4 = **12**	6. 7 ×2 = **14**
7. 7 ×5 = **35**	8. 1 ×4 = **4**	9. 3 ×5 = **15**	10. 2 ×2 = **4**	11. 8 ×3 = **24**	12. 4 ×3 = **12**
13. 4 ×6 = **24**	14. 5 ×2 = **10**	15. 4 ×5 = **20**	16. 2 ×9 = **18**	17. 5 ×5 = **25**	18. 5 ×6 = **30**
19. 4 ×2 = **8**	20. 0 ×9 = **0**	21. 9 ×3 = **27**	22. 4 ×4 = **16**	23. 3 ×7 = **21**	24. 8 ×2 = **16**
25. 6 ×2 = **12**	26. 3 ×6 = **18**	27. 5 ×4 = **20**	28. 4 ×7 = **28**	29. 3 ×2 = **6**	30. 9 ×5 = **45**
31. 5 ×1 = **5**	32. 2 ×0 = **0**	33. 3 ×1 = **3**	34. 4 ×9 = **36**	35. 5 ×0 = **0**	36. 3 ×8 = **24**

Answer Key

Name _____ (3.OA.C.7)

Multiplication and Division Fluency

Solve each problem.

1. $8 \times 5 = 40$	2. $6 \times 4 = 24$	3. $5 \times 5 = 25$	4. $9 \times 5 = 45$	5. $4 \times 4 = 16$	6. $6 \times 3 = 18$
7. $7 \times 4 = 28$	8. $7 \times 3 = 21$	9. $3 \times 8 = 24$	10. $6 \times 2 = 12$	11. $9 \times 3 = 27$	12. $5 \times 3 = 15$
13. $5 \times 4 = 20$	14. $8 \times 8 = 64$	15. $5 \times 6 = 30$	16. $7 \times 6 = 42$	17. $9 \times 8 = 72$	18. $9 \times 7 = 63$
19. $5 \overline{)15} = 3$	20. $1 \overline{)9} = 9$	21. $3 \overline{)27} = 9$	22. $8 \overline{)64} = 8$	23. $6 \overline{)24} = 4$	24. $6 \overline{)36} = 6$
25. $8 \overline{)32} = 4$	26. $4 \overline{)36} = 9$	27. $8 \overline{)24} = 3$	28. $7 \overline{)14} = 2$	29. $2 \overline{)18} = 9$	30. $8 \overline{)8} = 1$

Name _____ (3.OA.C.7)

Multiplication and Division Fluency

Solve each problem.

1. $2 \times 7 = 14$	2. $7 \times 5 = 35$	3. $5 \times 6 = 30$	4. $2 \times 7 = 14$	5. $4 \times 8 = 32$	6. $7 \times 6 = 42$
7. $4 \times 9 = 36$	8. $7 \times 3 = 21$	9. $6 \times 6 = 36$	10. $5 \times 9 = 45$	11. $4 \times 2 = 8$	12. $2 \times 6 = 12$
13. $3 \times 6 = 18$	14. $8 \times 7 = 56$	15. $8 \times 6 = 48$	16. $5 \times 7 = 35$	17. $1 \times 5 = 5$	18. $4 \times 5 = 20$
19. $8 \times 5 = 40$	20. $4 \times 3 = 12$	21. $6 \times 5 = 30$	22. $7 \times 7 = 49$	23. $9 \times 6 = 54$	24. $7 \times 9 = 63$
25. $8 \overline{)64} = 8$	26. $9 \overline{)18} = 2$	27. $6 \overline{)48} = 8$	28. $6 \overline{)54} = 9$	29. $8 \overline{)56} = 7$	30. $4 \overline{)32} = 8$
31. $3 \overline{)24} = 8$	32. $2 \overline{)14} = 7$	33. $8 \overline{)16} = 2$	34. $5 \overline{)45} = 9$	35. $4 \overline{)20} = 5$	36. $9 \overline{)81} = 9$

Name _____ (3.OA.C.7)

Multiplication and Division Fluency

Solve each problem.

1. $9 \times 4 = 36$	2. $6 \times 9 = 54$	3. $5 \times 6 = 30$	4. $5 \times 3 = 15$	5. $7 \times 9 = 63$	6. $8 \times 3 = 24$
7. $6 \times 7 = 42$	8. $4 \times 5 = 20$	9. $4 \times 2 = 8$	10. $4 \times 4 = 16$	11. $9 \times 7 = 63$	12. $8 \times 8 = 64$
13. $7 \times 2 = 14$	14. $8 \times 6 = 48$	15. $6 \times 8 = 48$	16. $4 \times 6 = 24$	17. $8 \times 4 = 32$	18. $6 \times 3 = 18$
19. $8 \times 5 = 40$	20. $9 \times 8 = 72$	21. $4 \times 3 = 12$	22. $8 \times 9 = 72$	23. $6 \times 5 = 30$	24. $9 \times 2 = 18$
25. $5 \overline{)30} = 6$	26. $4 \overline{)36} = 9$	27. $2 \overline{)18} = 9$	28. $4 \overline{)36} = 9$	29. $3 \overline{)27} = 9$	30. $9 \overline{)9} = 1$
31. $3 \overline{)24} = 8$	32. $4 \overline{)32} = 8$	33. $3 \overline{)9} = 3$	34. $8 \overline{)56} = 7$	35. $4 \overline{)36} = 9$	36. $8 \overline{)32} = 4$
37. $8 \overline{)64} = 8$	38. $9 \overline{)81} = 9$	39. $7 \overline{)28} = 4$	40. $7 \overline{)49} = 7$	41. $8 \overline{)16} = 2$	42. $1 \overline{)4} = 4$

Name _____ (3.OA.A.3, 3.OA.D.8)

Multistep Word Problems

> Anthony and his Scout troop hiked 2 miles and then rested. After their break, they hiked another 3 miles. The total distance of the hike was 7 miles. How many more miles did they need to go before reaching their destination?
>
> First: Add the total distance hiked so far. $2 + 3 = 5$
>
> Then: Subtract this sum from the total to find the remaining distance. $7 - 5 = 2$
>
> Answer: They must hike 2 more miles to reach their destination.

Read each problem. Show your work for both parts of the problem. Record your answer in the space given.

1. Our family drove 453 miles on vacation. We crossed 4 states. We stopped 2 times in 3 states and 3 times in the last state. How many times did we stop?
   First: $2 \times 3 = 6$
   Then: $6 + 3 = 9$
   We stopped **9** times.

2. We went to an amusement park. There were 4 roller coasters. Jill rode each one 3 times. Henry rode each one twice. How many times did Jill and Henry ride them in all?
   First: $4 \times 2 = 8$
   Then: $12 + 8 = 20$
   They rode **20** times.

3. We hiked 9 miles on both Monday and Tuesday. We rode our bikes 23 miles on Tuesday and 31 miles on Wednesday. How far did we travel altogether?
   First: $9 \times 2 = 18$
   Then: $18 + 23 + 31 = 72$
   We traveled **72** miles.

4. Mia reads books in the car. On one trip, she read 173 pages. On another trip, she read 194 pages. If the book is 421 pages long, how many pages does she have left to read?
   First: $173 + 194 = 367$
   Then: $421 - 367 = 54$
   She has **54** pages left.

5. Jared's goal is to find 1,000 insects. In one park, he counted 356 bugs. In another park, he counted 358 bugs. How many more does he need to find?
   First: $356 + 358 = 714$
   Then: $1000 - 714 = 286$
   He needs to find **286** more bugs.

6. Laura found 23 flowers in her backyard. She found 22 more in her friend's yard. She divided the flowers equally between 5 vases. How many flowers did she put in each vase?
   First: $23 + 22 = 45$
   Then: $45 \div 5 = 9$
   She put **9** flowers in each vase.

Answer Key

Name _____ 3.OA.A.3, 3.OA.D.8

Multistep Word Problems

Use two or more steps to solve each problem.

1. Lola and Mariah have each blown up 4 balloons. They need a total of 20 balloons for the party at school. How many more do they need to blow up?

$4 \times 2 =$ **8**

then, $20 -$ **8** $=$ **12**   They need **12** more balloons.

2. Lee has collected a total of 29 cans and boxes of food for the food drive. Chung has collected 27. Their team goal was to collect 75 cans and boxes of food altogether. How many more do they need to collect to reach their awesome goal?

29 + 27 = 56

75 − 56 = 19   **19 more cans and boxes**

3. Yesterday, Eric picked 34 apples off the tree in his backyard. Today, he picked another 66 apples. His father asked him to divide the total in half so that they could share their apples with the food bank. How many apples did they donate?

34 + 66 = 100

100 ÷ 2 = 50   **50 apples**

4. Roberto, Heather, and Macon each raised 3 dollars for the fund-raiser at school by selling pretzels at the carnival. They need a total of 15 dollars to reach their goal. How much more do they need to raise to reach their goal?

3 × 3 = 9

15 − 9 = 6   **$6 more**

5. Wyatt pays 2 dollars each time he goes to the water park to swim. He has already been there 5 times this summer! How many more times can he go before he spends a total of 20 dollars?

2 × 5 = 10

20 − 10 = 10   **10 ÷ 2 = 5**   **5 times**

6. Zaina loves to draw! She draws on 8 sheets of her pad a day. She's had her 80-page drawing pad for 9 days. How many clean pages does she have left?

8 × 9 = 72

80 − 72 = 8   **8 pages**

Name _____ 3.OA.A.3, 3.OA.D.8

Multistep Word Problems

Solve. Show your work.

1. Seth has 3 nickels in each of his 4 pockets. How much money does Seth have? **60¢**	2. Three bags each have 9 marbles. Ten of the marbles rolled under the table. How many are left? **17 marbles**
3. Four bags each hold 8 fruit snacks. Five snacks were eaten. Are there enough left to share equally with 3 people? Explain. **Yes, 27 ÷ 3 = 9 people**	4. Four cups are on the table. Each cup needs 6 gumdrops in it. One bag has 25 gumdrops in it. Is one bag enough to fill the 4 cups? Explain. **Yes, 6 × 4 = 24 gumdrops**
5. Victor practiced his free throws. On 3 days, he made 9 shots each day. On 4 days, he made 7 shots each day. How many shots did he make all week? **55 shots**	6. Six words each have 2 vowels and 2 consonants. How many letters are there in the 6 words? How many vowels? Consonants? **24 letters, 12 vowels, 12 consonants**
7. Seven bags each hold 9 unit cubes. Twenty-one more cubes were found. What is the total number of cubes? If those 21 cubes were put evenly into the original 7 bags, how many unit cubes would there be in each bag? **84 total cubes, 12 cubes in each bag**	8. Pilar used 253 beads to make a bracelet. She used 704 beads to make a necklace. If the package started with 1,000 beads, how many beads are left? **43 beads**

Check students' work.

Name _____ 3.OA.D.9

Arithmetic Patterns

Multiply to complete the chart.

×	0	1	2	3	4	5	6	7	8	9	10
0	0	0	0	0	0	0	0	0	0	0	0
1	0	1	2	3	4	5	6	7	8	9	10
2	0	2	4	6	8	10	12	14	16	18	20
3	0	3	6	9	12	15	18	21	24	27	30
4	0	4	8	12	16	20	24	28	32	36	40
5	0	5	10	15	20	25	30	35	40	45	50
6	0	6	12	18	24	30	36	42	48	54	60
7	0	7	14	21	28	35	42	49	56	63	70
8	0	8	16	24	32	40	48	56	64	72	80
9	0	9	18	27	36	45	54	63	72	81	90
10	0	10	20	30	40	50	60	70	80	90	100

1. Color the product of double factors red.

2. Complete each sentence.

Multiples of zero always equal ___**0**___.

Multiples of ten always end with ___**0**___.

Multiples of five always end with ___**0**___ or ___**5**___.

3. Write even or odd in each blank.

even number × even number = ___**even**___ number

odd number × odd number = ___**odd**___ number

even number × odd number = ___**even**___ number

Name _____ 3.OA.D.9

Arithmetic Patterns

Fill in the answers on the times table below.

×	0	1	2	3	4	5	6	7	8	9	10
0	0	0	0	0	0	0	0	0	0	0	0
1	0	1	2	3	4	5	6	7	8	9	10
2	0	2	4	6	8	10	12	14	16	18	20
3	0	3	6	9	12	15	18	21	24	27	30
4	0	4	8	12	16	20	24	28	32	36	40
5	0	5	10	15	20	25	30	35	40	45	50
6	0	6	12	18	24	30	36	42	48	54	60
7	0	7	14	21	28	35	42	49	56	63	70
8	0	8	16	24	32	40	48	56	64	72	80
9	0	9	18	27	36	45	54	63	72	81	90
10	0	10	20	30	40	50	60	70	80	90	100

Circle odd, even, or both to describe the patterns in the table.

1. Count by ones.   odd, even, (both)
2. Count by twos.   odd, (even), both
3. Count by threes.   odd, even, (both)
4. Count by fours.   odd, (even), both
5. Count by fives.   odd, even, (both)
6. Count by sixes.   odd, (even), both
7. Count by sevens.   odd, even, (both)
8. Count by eights.   odd, (even), both
9. Count by nines.   odd, even, (both)
10. Count by tens.   odd, (even), both

11. Write even or odd in each blank.

even number × even number = ___**even**___ number

odd number × odd number = ___**odd**___ number

even number × odd number = ___**even**___ number

Answer Key

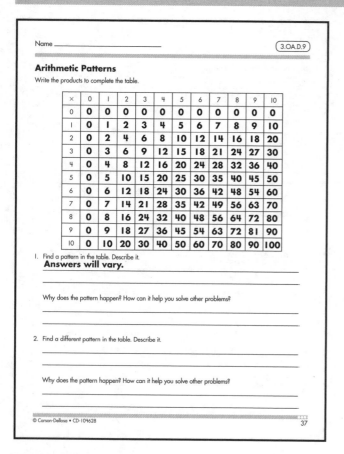

3.OA.D.9

Arithmetic Patterns

Write the products to complete the table.

×	0	1	2	3	4	5	6	7	8	9	10
0	0	0	0	0	0	0	0	0	0	0	0
1	0	1	2	3	4	5	6	7	8	9	10
2	0	2	4	6	8	10	12	14	16	18	20
3	0	3	6	9	12	15	18	21	24	27	30
4	0	4	8	12	16	20	24	28	32	36	40
5	0	5	10	15	20	25	30	35	40	45	50
6	0	6	12	18	24	30	36	42	48	54	60
7	0	7	14	21	28	35	42	49	56	63	70
8	0	8	16	24	32	40	48	56	64	72	80
9	0	9	18	27	36	45	54	63	72	81	90
10	0	10	20	30	40	50	60	70	80	90	100

1. Find a pattern in the table. Describe it.
Answers will vary.

Why does the pattern happen? How can it help you solve other problems?

2. Find a different pattern in the table. Describe it.

Why does the pattern happen? How can it help you solve other problems?

© Carson-Dellosa • CD-104628 — 37

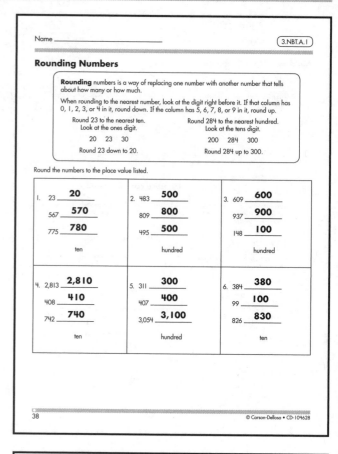

3.NBT.A.1

Rounding Numbers

Rounding numbers is a way of replacing one number with another number that tells about how many or how much.

When rounding to the nearest number, look at the digit right before it. If that column has 0, 1, 2, 3, or 4 in it, round down. If the column has 5, 6, 7, 8, or 9 in it, round up.

Round 23 to the nearest ten.
Look at the ones digit.

20   23   30

Round 23 down to 20.

Round 284 to the nearest hundred.
Look at the tens digit.

200   284   300

Round 284 up to 300.

Round the numbers to the place value listed.

1. 23	**20**	2. 483	**500**	3. 609	**600**
567	**570**	809	**800**	937	**900**
775	**780**	495	**500**	148	**100**
	ten		hundred		hundred
4. 2,813	**2,810**	5. 311	**300**	6. 384	**380**
408	**410**	407	**400**	99	**100**
742	**740**	3,054	**3,100**	826	**830**
	ten		hundred		ten

38 — © Carson-Dellosa • CD-104628

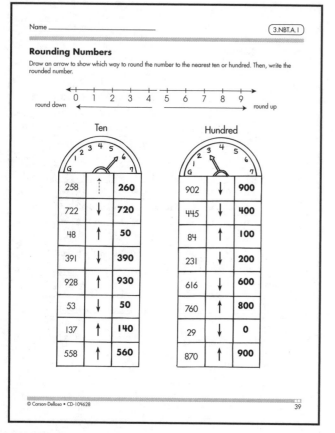

3.NBT.A.1

Rounding Numbers

Draw an arrow to show which way to round the number to the nearest ten or hundred. Then, write the rounded number.

round down ← 0 1 2 3 4 | 5 6 7 8 9 → round up

Ten

258	↑	**260**
722	↓	**720**
48	↑	**50**
391	↓	**390**
928	↑	**930**
53	↓	**50**
137	↑	**140**
558	↑	**560**

Hundred

902	↓	**900**
445	↓	**400**
84	↑	**100**
231	↓	**200**
616	↓	**600**
760	↑	**800**
29	↓	**0**
870	↑	**900**

© Carson-Dellosa • CD-104628 — 39

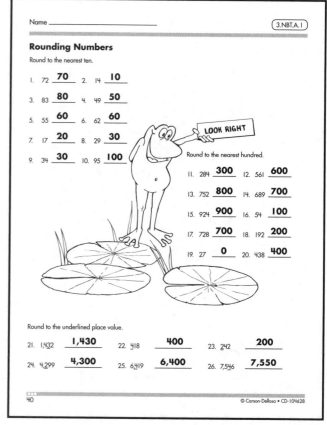

3.NBT.A.1

Rounding Numbers

Round to the nearest ten.

1. 72 **70**   2. 14 **10**
3. 83 **80**   4. 49 **50**
5. 55 **60**   6. 62 **60**
7. 17 **20**   8. 29 **30**
9. 34 **30**   10. 95 **100**

LOOK RIGHT

Round to the nearest hundred.

11. 284 **300**   12. 561 **600**
13. 752 **800**   14. 689 **700**
15. 924 **900**   16. 54 **100**
17. 728 **700**   18. 192 **200**
19. 27 **0**   20. 438 **400**

Round to the underlined place value.

21. 1,4̲32 **1,430**   22. 4̲18 **400**   23. 2̲42 **200**
24. 4,2̲99 **4,300**   25. 6,4̲19 **6,400**   26. 7,5̲46 **7,550**

40 — © Carson-Dellosa • CD-104628

© Carson-Dellosa • CD-104628

Answer Key

Page 41

Addition and Subtraction within 1,000

When the sum in the tens column is greater than 9, regroup to the hundreds column. It looks like this:

1. Add the ones. Regroup if needed.
   172
   + 473
   ___5

2. Add the tens. Regroup if needed.
   1
   172
   + 473
   __45

3. Add the hundreds.
   172
   + 473
   _645

When the top number in the tens column is less than the bottom number, you must regroup from the hundreds. It looks like this:

1. Subtract the ones. Regroup if needed.
   607
   − 284
   ___3

2. Subtract the tens. Regroup if needed.
   5 10
   6Ø7
   − 284
   __23

3. Subtract the hundreds.
   5 10
   6Ø7
   − 284
   _323

Solve each problem. Regroup when necessary.

1. 634 + 268 = **902**
2. 987 + 489 = **1,476**
3. 768 − 479 = **289**
4. 888 + 276 = **1,164**
5. 747 − 458 = **289**
6. 950 − 580 = **370**

7. 394 + 496 = **890**
8. 689 − 478 = **211**
9. 254 + 347 = **601**
10. 665 + 337 = **1,002**
11. 521 − 295 = **226**
12. 988 + 748 = **1,736**

13. 301 − 242 = **59**
14. 349 + 233 = **582**
15. 878 + 287 = **1,165**
16. 727 − 49 = **678**
17. 348 + 948 = **1,296**
18. 847 − 358 = **489**

19. 477 + 298 = **775**
20. 704 − 597 = **107**
21. 846 − 457 = **389**
22. 834 + 249 = **1,083**
23. 405 − 228 = **177**
24. 118 + 953 = **1,071**

25. 703 − 478 = **225**
26. 653 + 307 = **960**
27. 584 − 295 = **289**
28. 600 − 367 = **233**
29. 113 + 298 = **411**
30. 393 + 298 = **691**

Page 42

Addition and Subtraction within 1,000

Solve each problem. Regroup when necessary.

1. 784 − 591 = **193**
2. 979 + 654 = **1,633**
3. 945 + 379 = **1,324**
4. 825 − 638 = **187**
5. 870 + 739 = **1,609**
6. 478 + 655 = **1,133**

7. 675 + 597 = **1,272**
8. 456 + 327 = **783**
9. 654 − 265 = **389**
10. 293 − 187 = **106**
11. 765 + 428 = **1,193**
12. 824 − 548 = **276**

13. 845 − 566 = **279**
14. 349 + 233 = **582**
15. 434 + 948 = **1,382**
16. 725 − 469 = **256**
17. 827 − 529 = **298**
18. 638 + 422 = **1,060**

19. 539 + 468 = **1,007**
20. 574 − 293 = **281**
21. 955 + 134 = **1,089**
22. 423 − 155 = **268**
23. 588 + 294 = **882**
24. 536 − 258 = **278**

25. 666 + 291 = **957**
26. 857 − 675 = **182**
27. 646 + 668 = **1,314**
28. 748 − 353 = **395**
29. 831 − 357 = **474**
30. 348 + 489 = **837**

31. 239 + 293 = **532**
32. 438 − 254 = **184**
33. 487 − 391 = **96**
34. 159 + 485 = **644**
35. 648 + 437 = **1,085**
36. 940 − 556 = **384**

Page 43

Addition and Subtraction within 1,000

Solve each problem. Regroup when necessary.

1. 539 − 375 = **164**
2. 476 + 243 = **719**
3. 176 + 484 = **660**
4. 392 + 292 = **684**
5. 787 − 598 = **189**
6. 165 + 427 = **592**

7. 481 + 428 = **909**
8. 842 + 177 = **1,019**
9. 762 − 395 = **367**
10. 856 − 399 = **457**
11. 347 + 983 = **1,330**
12. 275 + 298 = **573**

13. 628 − 127 = **501**
14. 531 − 467 = **64**
15. 496 − 288 = **208**
16. 389 + 392 = **781**
17. 276 + 391 = **667**
18. 374 − 276 = **98**

19. 983 − 468 = **515**
20. 834 − 376 = **458**
21. 452 + 287 = **739**
22. 392 + 284 = **676**
23. 597 − 387 = **210**
24. 584 − 287 = **297**

25. 498 − 268 = **230**
26. 735 + 373 = **1,108**
27. 395 + 161 = **556**
28. 502 − 321 = **181**
29. 890 − 249 = **641**
30. 848 − 399 = **449**

31. 400 − 373 = **27**
32. 837 − 209 = **628**
33. 622 − 323 = **299**
34. 786 − 389 = **397**
35. 663 − 261 = **402**
36. 429 − 188 = **241**

Page 44

Multiplying by 10s

Follow these steps to multiply with a multiple of 10.

1. Multiply the nonzero digits.
   30
   × 4
   __12

2. Add a 0. Because the top number you multiplied by is 10 times larger, the product is too.
   30
   × 4
   _120

Multiply. The first one is done for you.

1. 90 × 2 = **180**
2. 60 × 3 = **180**
3. 80 × 8 = **640**
4. 40 × 3 = **120**

5. 60 × 9 = **540**
6. 70 × 2 = **140**
7. 90 × 7 = **630**
8. 50 × 4 = **200**

9. 70 × 3 = **210**
10. 60 × 6 = **360**
11. 50 × 2 = **100**
12. 70 × 5 = **350**

13. 20 × 4 = **80**
14. 20 × 3 = **60**
15. 30 × 8 = **240**
16. 20 × 5 = **100**

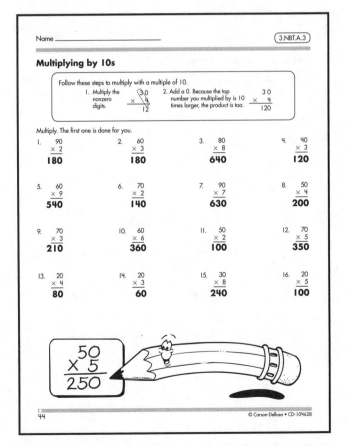

50
× 5
___
250

Answer Key

Name _____   3.NBT.A.3

Multiplying by 10s

Multiply. The first one is done for you.

1. 90
    × 2
    180

2. 60
    × 5
    300

3. 70
    × 4
    280

4. 50
    × 7
    350

5. 60
    × 9
    540

6. 50
    × 3
    150

7. 20
    × 6
    120

8. 60
    × 8
    480

9. 50
    × 8
    400

10. 30
    × 7
    210

11. 20
    × 9
    180

12. 80
    × 6
    480

13. 90
    × 9
    810

14. 70
    × 6
    420

15. 80
    × 8
    640

16. 30
    × 8
    240

45

Name _____   3.NBT.A.3

Multiplying by 10s

Multiply.

1. 20
    × 3
    60

2. 20
    × 4
    80

3. 40
    × 2
    80

4. 10
    × 7
    70

5. 10
    × 5
    50

6. 20
    × 9
    180

7. 90
    × 3
    270

8. 40
    × 6
    240

9. 60
    × 4
    240

10. 90
    × 9
    810

11. 60
    × 7
    420

12. 30
    × 5
    150

13. 30
    × 8
    240

14. 40
    × 3
    120

15. 70
    × 5
    350

16. 80
    × 6
    480

17. 90
    × 2
    180

18. 30
    × 4
    120

19. 40
    × 6
    240

20. 80
    × 3
    240

21. 70
    × 7
    490

22. 90
    × 4
    360

23. 70
    × 8
    560

24. 50
    × 8
    400

46

Name _____   3.NF.A.1

Understanding Unit Fractions

This circle has been divided into equal parts, so it can also be described as a fraction.
Example:

■ = $\frac{1}{4}$ ← ■ parts
          ← total parts

Use the circles to answer the questions.

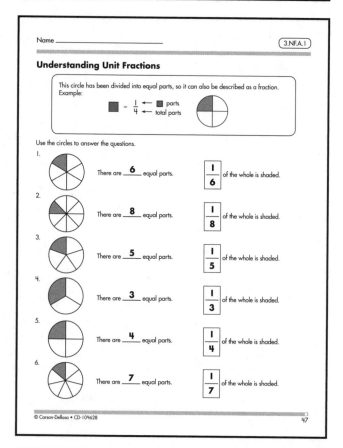

1. There are __6__ equal parts. $\frac{1}{6}$ of the whole is shaded.

2. There are __8__ equal parts. $\frac{1}{8}$ of the whole is shaded.

3. There are __5__ equal parts. $\frac{1}{5}$ of the whole is shaded.

4. There are __3__ equal parts. $\frac{1}{3}$ of the whole is shaded.

5. There are __4__ equal parts. $\frac{1}{4}$ of the whole is shaded.

6. There are __7__ equal parts. $\frac{1}{7}$ of the whole is shaded.

47

Name _____   3.NF.A.1

Understanding Unit Fractions

Shade and label the unit fraction for each whole. Some fractions may be used more than once.

$\frac{1}{2}$ $\frac{1}{3}$ $\frac{1}{4}$ $\frac{1}{6}$ $\frac{1}{8}$

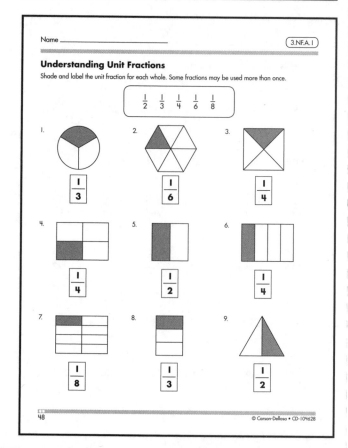

1. $\frac{1}{3}$

2. $\frac{1}{6}$

3. $\frac{1}{4}$

4. $\frac{1}{4}$

5. $\frac{1}{2}$

6. $\frac{1}{4}$

7. $\frac{1}{8}$

8. $\frac{1}{3}$

9. $\frac{1}{2}$

48

Answer Key

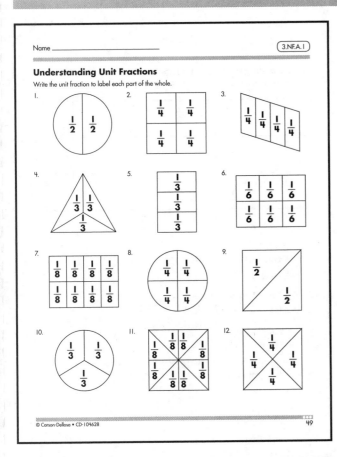

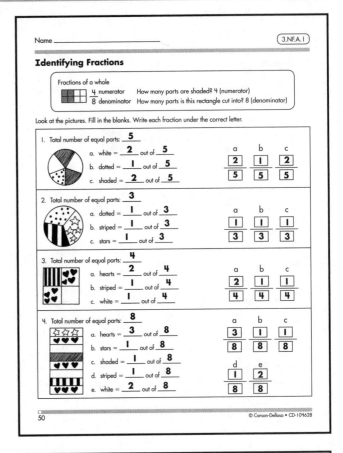

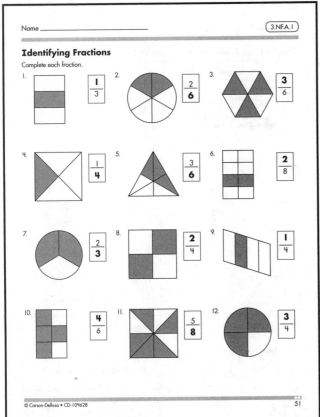

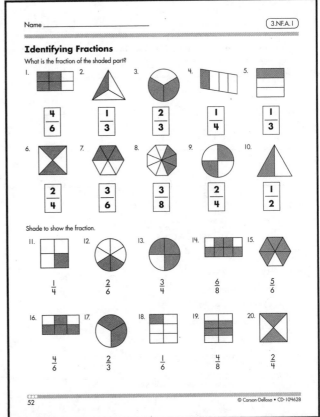

Answer Key

Name _____ (3.NF.A.2)

Understanding Fractions on a Number Line

Fractions can be represented on a number line.

Both the number line and the rectangle represent the fraction $\frac{1}{3}$.

Use the number line to complete the questions.

1. The number line is divided into **4** pieces.

2. Label the number line with the correct fractions.

3. Each section of the number line represents what fraction? **$\frac{1}{4}$**

4. Draw a dot to show the location of $\frac{2}{4}$.

5. The number line is divided into **6** pieces.

6. Label the number line with the correct fractions.

7. Each section of the number line represents what fraction? **$\frac{1}{6}$**

8. Draw a dot to show the location of $\frac{5}{6}$.

© Carson-Dellosa • CD-104628   53

Name _____ (3.NF.A.2)

Understanding Fractions on a Number Line

Use the number lines to answer the questions.

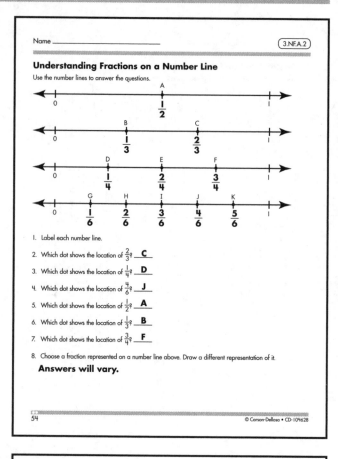

1. Label each number line.

2. Which dot shows the location of $\frac{2}{3}$? **C**

3. Which dot shows the location of $\frac{1}{4}$? **D**

4. Which dot shows the location of $\frac{4}{6}$? **J**

5. Which dot shows the location of $\frac{1}{2}$? **A**

6. Which dot shows the location of $\frac{1}{3}$? **B**

7. Which dot shows the location of $\frac{3}{4}$? **F**

8. Choose a fraction represented on a number line above. Draw a different representation of it.

Answers will vary.

54   © Carson-Dellosa • CD-104628

Name _____ (3.NF.A.2, 3.NF.A.3c)

Understanding Fractions on a Number Line

Use the number lines to answer the questions.

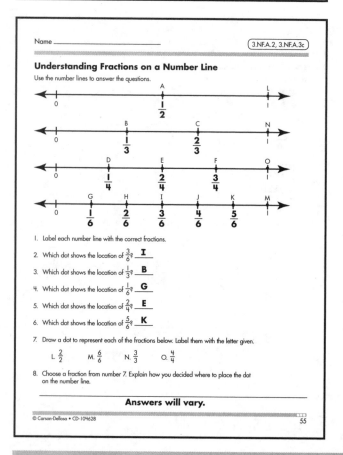

1. Label each number line with the correct fractions.

2. Which dot shows the location of $\frac{3}{6}$? **I**

3. Which dot shows the location of $\frac{1}{3}$? **B**

4. Which dot shows the location of $\frac{1}{6}$? **G**

5. Which dot shows the location of $\frac{2}{4}$? **E**

6. Which dot shows the location of $\frac{5}{6}$? **K**

7. Draw a dot to represent each of the fractions below. Label them with the letter given.

   L. $\frac{2}{2}$ M. $\frac{6}{6}$ N. $\frac{3}{3}$ O. $\frac{4}{4}$

8. Choose a fraction from number 7. Explain how you decided where to place the dot on the number line.

Answers will vary.

© Carson-Dellosa • CD-104628   55

Name _____ (3.NF.A.2)

Representing Fractions on a Number Line

To show a fraction on a number line, draw a number line with 0 and 1 as the endpoints.

Then, divide it into equal parts and label them.

Finally, draw a dot on the line to represent the fraction.

Show each fraction on the number line.

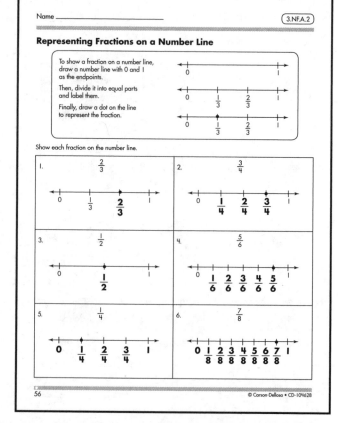

56   © Carson-Dellosa • CD-104628

Answer Key

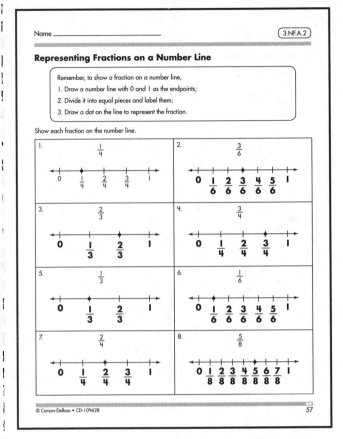

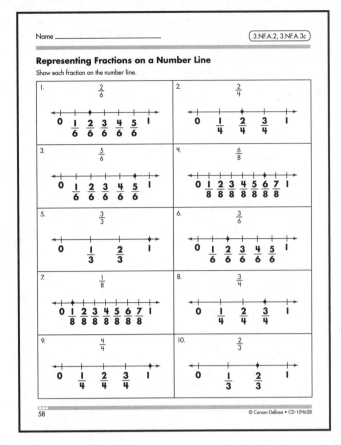

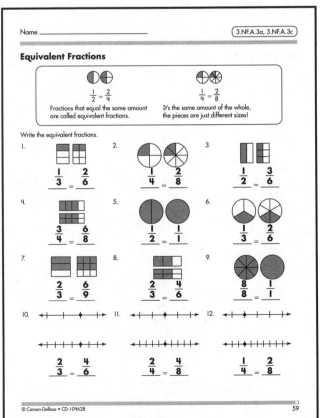

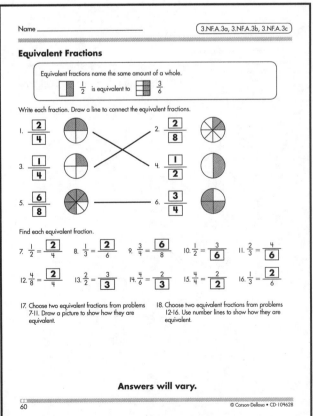

Answer Key

Name _____

3.NF.A.3a, 3.NF.A.3b, 3.NF.A.3c

Equivalent Fractions

Decide if each pair of fractions is equivalent. Draw an X over the pairs that are not equivalent. Draw a picture to explain why or why not.

1. $\frac{1}{4}$ and $\frac{3}{8}$ ✗
2. $\frac{2}{3}$ and $\frac{2}{4}$ ✗
3. $\frac{4}{8}$ and $\frac{1}{2}$

4. $\frac{2}{2}$ and $\frac{6}{6}$
5. $\frac{1}{2}$ and $\frac{3}{6}$
6. $\frac{6}{8}$ and $\frac{4}{...}$ ✗

7. $\frac{3}{4}$ and $\frac{2}{3}$ ✗
8. $\frac{3}{3}$ and $\frac{4}{4}$
9. $\frac{2}{3}$ and $\frac{3}{6}$ ✗

Check students' drawings.

Name _____

3.NF.A.3d

Comparing Fractions

When comparing fractions, look at the numerators and denominators.

If the **numerators** are the same, compare the denominators. The fraction with the smaller denominator is divided into fewer, larger pieces. So it is the greater fraction.

If the **denominators** are the same, compare the numerators. The fraction with the bigger numerator has more of the same-size pieces. So it is the greater fraction.

Identify each fraction. Circle the greater fraction.

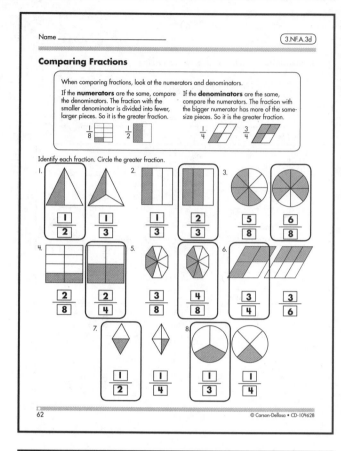

Name _____

3.NF.A.3d

Comparing Fractions

Circle the comparisons in each set that are not true. Rewrite the false comparisons so that they are true. Draw pictures to prove your corrections.

1. $\frac{1}{4} < \frac{1}{3}$ $\frac{1}{6} > \frac{1}{9}$ ⬭$\frac{1}{5} > \frac{1}{2}$⬭   ⬭$\frac{1}{3} < \frac{1}{5}$⬭ $\frac{1}{7} > \frac{1}{10}$
   $$\frac{1}{5} < \frac{1}{2} \qquad \frac{1}{3} > \frac{1}{5}$$

2. ⬭$\frac{2}{5} > \frac{4}{5}$⬭ $\frac{6}{7} > \frac{2}{7}$ $\frac{1}{3} < \frac{2}{3}$ ⬭$\frac{4}{8} > \frac{6}{8}$⬭ $\frac{1}{9} < \frac{4}{9}$
   $$\frac{2}{5} < \frac{4}{5} \qquad\qquad\qquad \frac{4}{8} < \frac{6}{8}$$

3. ⬭$\frac{1}{2} < \frac{1}{4}$⬭ $\frac{3}{4} > \frac{1}{4}$ ⬭$\frac{5}{6} > \frac{2}{6}$⬭ $\frac{1}{5} > \frac{1}{10}$ $\frac{4}{5} > \frac{2}{5}$
   $$\frac{1}{2} > \frac{1}{4} \qquad\qquad\qquad \frac{5}{6} > \frac{2}{6}$$

4. $\frac{5}{8} > \frac{2}{8}$ ⬭$\frac{1}{6} < \frac{1}{10}$⬭   ⬭$\frac{2}{9} > \frac{4}{9}$⬭ $\frac{1}{7} > \frac{1}{12}$ $\frac{1}{2} > \frac{1}{11}$
   $$\frac{1}{6} > \frac{1}{10} \qquad \frac{2}{9} < \frac{4}{9}$$

5. $\frac{4}{5} > \frac{3}{5}$ $\frac{2}{6} < \frac{5}{6}$ ⬭$\frac{1}{3} < \frac{1}{6}$⬭   ⬭$\frac{1}{4} > \frac{2}{4}$⬭ $\frac{1}{5} > \frac{1}{8}$
   $$\frac{1}{3} > \frac{1}{6} \qquad \frac{1}{4} < \frac{2}{4}$$

6. $\frac{1}{7} > \frac{1}{8}$ ⬭$\frac{1}{8} > \frac{1}{4}$⬭ $\frac{7}{9} > \frac{3}{9}$ $\frac{4}{10} < \frac{7}{10}$ ⬭$\frac{5}{6} < \frac{3}{6}$⬭
   $$\frac{1}{8} < \frac{1}{4} \qquad\qquad\qquad\qquad \frac{5}{6} > \frac{3}{6}$$

Name _____

3.NF.A.3d

Comparing Fractions

Compare using $>$, $<$, or $=$.

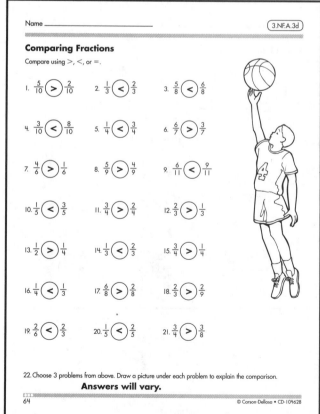

1. $\frac{5}{10} > \frac{2}{10}$ 2. $\frac{1}{3} < \frac{2}{3}$ 3. $\frac{5}{8} < \frac{6}{8}$

4. $\frac{3}{10} < \frac{8}{10}$ 5. $\frac{1}{4} < \frac{3}{4}$ 6. $\frac{6}{7} > \frac{3}{7}$

7. $\frac{4}{6} > \frac{1}{6}$ 8. $\frac{5}{9} > \frac{1}{9}$ 9. $\frac{6}{11} < \frac{9}{11}$

10. $\frac{1}{5} < \frac{3}{5}$ 11. $\frac{3}{4} > \frac{2}{4}$ 12. $\frac{2}{3} > \frac{1}{3}$

13. $\frac{1}{2} > \frac{1}{4}$ 14. $\frac{1}{3} < \frac{2}{3}$ 15. $\frac{3}{4} > \frac{1}{4}$

16. $\frac{1}{4} < \frac{1}{3}$ 17. $\frac{6}{8} > \frac{2}{8}$ 18. $\frac{2}{3} > \frac{2}{9}$

19. $\frac{2}{6} < \frac{2}{3}$ 20. $\frac{1}{5} < \frac{2}{5}$ 21. $\frac{3}{4} > \frac{3}{8}$

22. Choose 3 problems from above. Draw a picture under each problem to explain the comparison.
Answers will vary.

Answer Key

Elapsed Time

To calculate elapsed time, or time that has passed, try these methods:

The school play started at 3:00. It lasted 2 hours 10 minutes. What time did it end?

Step 1: Start at 3:00. Add or count the hours.

Step 2: Add or count the minutes.

It will end at 5:10.

Step 1: Draw a number line that marks the starting and ending times. Write the start and end times if you know them.

Step 2: Draw a line to show the amount of time to get from the start time to the next hour, and to get to the previous hour from the end time. Write how much time each space covers.

Step 3: Mark how many hours are represented in the middle.

Step 4: Add the times together.

Use the clocks to help you find the elapsed time.

1. Harry rode his bike from 4:00 to 4:58. How long did he ride his bike? **58 minutes**

2. Jeff's game started at 2:00. It ended at 3:43. How long did the game last? **1 hour 43 minutes**

3. Zoe's favorite movie starts at 7:15. It will last for 2 hours and 18 minutes. What time will the movie end? **9:33**

4. Pollo's cat disappeared at 3:12. Pollo found him at 4:26. How long was his cat lost? **1 hour 14 minutes**

5. The train leaves for New York at 8:05. The ride is 3 hours 36 minutes long. What time will the train arrive in New York? **11:41**

6. Dawn started reading at 4:27. She read until 6:06. How long did Dawn read? **1 hour 39 minutes**

Elapsed Time

Use the clocks to help you find the elapsed time.

1. The dancing dogs came on at 6:15. They danced until 6:42. How long did they dance? **27 minutes**

2. The clowns started at 7:03. They rode bikes for 47 minutes. What time did they end? **7:50**

3. The lion tamer started at 4:21. He was on stage until 5:29. How long did he perform? **1 hour 8 minutes**

4. The elephants came on at 5:19. They performed for 1 hour and 4 minutes. What time did they finish? **6:23**

5. The tightrope walker began at 8:44. She finished at 9:28. How long was she walking? **44 minutes**

6. The human rocket came on at 9:45. He shot off the stage 13 minutes later. What time did he leave? **9:58**

7. Gabe's family left their house at 5:27 for the show. They sat in their seats at 5:56. How long did it take them to get to the circus? **29 minutes**

8. The show started at 6:02. It finally ended at 9:59. How long did the entire show take? **3 hours 57 minutes**

Elapsed Time

Read each problem. Use clocks or number lines to solve.

1. It is 11:54. The class will return from lunch at 12:25. How many minutes until the class returns? **31 minutes**

2. It is 7:11. The movie starts at 8:30. How long until the movie starts? **1 hour 19 minutes**

3. It is 4:43. Dinner is at 5:15. How long until dinner? **32 minutes**

4. The pizza takes 35 minutes to bake. I put it in the oven at 5:18. When will it be done? **5:53**

5. The soccer field is 26 minutes away. If we leave at 3:53, when do we arrive at the field? **4:19**

6. Jack's new game shuts down after 45 minutes. He started the game at 4:48. When will the game turn itself off? **5:33**

7. Room 12 goes to music at 2:35. Music is done at 3:15. How long is music? **40 minutes**

8. The video is 49 minutes long. We start watching it at 7:22. What time will the video be done? **8:11**

9. It is 8:07. The bus comes at 8:30. How long until the bus comes? **23 minutes**

10. Sheila ran 2 miles in 16 minutes. She started at 2:49. When did she finish? **3:05**

Mass and Liquid Volume

Metric units of mass
1,000 grams = 1 kilogram
1,000 g = 1 kg
1 kilogram (kg)
1 gram (g)

Metric units of capacity
1,000 millimeters = 1 liter
1,000 mL = 1 L
1 mL
1 L

Circle the correct unit to measure the following items.

1. **g** / kg

2. mL / **L**

3. **g** / kg

4. **g** / kg

5. mL / **L**

6. **mL** / L

7. **g** / kg

8. Mrs. Murphy filled a bucket to mop the floor. Does her bucket probably hold 10 milliliters or 10 liters of water? **10 liters**

9. A party hat has a mass of 30 grams. What is the mass of a set of 8 party hats? **240 grams**

10. Tony has a small pool that holds 150 liters of water. He has filled it with 87 liters of water so far. How many more liters can he add to the pool? **63 liters**

Answer Key

Name _____ 3.MD.A.2

Mass and Liquid Volume

Circle the best estimate.

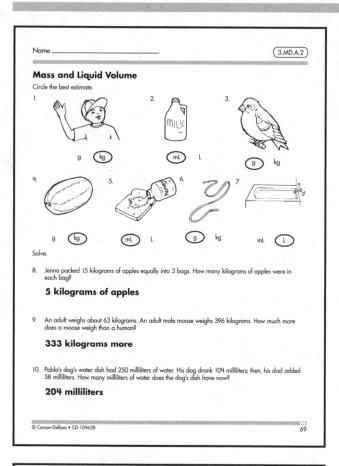

1. g (kg)
2. (mL) L
3. g kg

4. g (kg)
5. (mL) L
6. g kg
7. mL (L)

Solve.

8. Jenna packed 15 kilograms of apples equally into 3 bags. How many kilograms of apples were in each bag?

   5 kilograms of apples

9. An adult weighs about 63 kilograms. An adult male moose weighs 396 kilograms. How much more does a moose weigh than a human?

   333 kilograms more

10. Pablo's dog's water dish had 250 milliliters of water. His dog drank 104 milliliters; then, his dad added 58 milliliters. How many milliliters of water does the dog's dish have now?

    204 milliliters

© Carson-Dellosa • CD-104628   69

Name _____ 3.MD.A.2

Mass and Liquid Volume

Choose the best unit to measure each item (g, kg, mL, or L).

1. a bathtub **L**
2. a medicine dropper **mL**
3. a bird **g**
4. a bowl of soup **mL**
5. a baseball **g**
6. a large bottle of juice **L**
7. a golf cart **kg**
8. a cupcake **g**
9. a hat **g**
10. a pen **g**

11. Mr. Diaz bought drinks for a party. Did he buy 14 liters or 14 milliliters of drinks? How do you know?

    14 liters, Answers will vary.

12. Taylor, Quan, and India each have a book bag that weighs 4 kilograms. How many kilograms do their book bags weigh in all?

    12 kilograms

13. Hunter has a collection of bouncy balls. Each ball weighs 9 grams. His whole collection weighs 450 grams. How many bouncy balls are in Hunter's collection?

    50 bouncy balls

14. The grocery store stocked 2-liter bottles on the shelves. Each shelf can hold 8 bottles. There are 5 shelves in the section. How many liters can the store display at one time?

    80 liters

15. Lily prepared 234 milliliters of soup. Her brother made himself 281 milliliters of soup. If the pot started with 600 milliliters of soup, is there enough left for Lily's mom to have 250 milliliters? Why or why not?

    No, there are only 85 milliliters left.

70   © Carson-Dellosa • CD-104628

Name _____ 3.MD.B.3

Bar Graphs and Pictographs

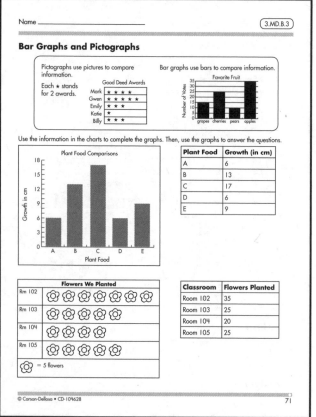

Pictographs use pictures to compare information.

Each ★ stands for 2 awards.

Good Deed Awards

Mark	★ ★ ★ ★
Gwen	★ ★ ★ ★ ★
Emily	★ ★ ★
Katie	★
Billy	★ ★ ★

Bar graphs use bars to compare information.

Use the information in the charts to complete the graphs. Then, use the graphs to answer the questions.

Plant Food	Growth (in cm)
A	6
B	13
C	17
D	6
E	9

Flowers We Planted

Classroom	Flowers Planted
Room 102	35
Room 103	25
Room 104	20
Room 105	25

= 5 flowers

© Carson-Dellosa • CD-104628   71

Name _____ 3.MD.B.3

Bar Graphs and Pictographs

Use the information in the charts to complete the graphs. Then, use the graphs to answer the questions.

1. How many more students voted for sea lions than birds? **9 students**
2. How many more students voted for snakes than elephants? **6 students**
3. How many votes did elephants and birds receive altogether? **18 votes**
4. How many votes did sea lions and snakes receive altogether? **33 votes**
5. How many votes were recorded altogether? **51 votes**

Animal	Votes
elephant	12
sea lion	15
bird	6
snake	18

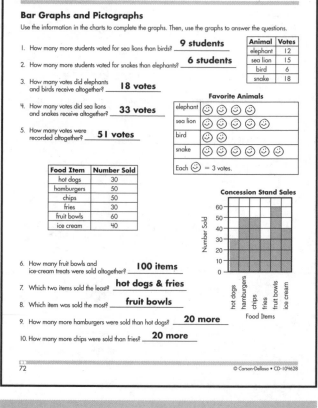

Food Item	Number Sold
hot dogs	30
hamburgers	50
chips	50
fries	30
fruit bowls	60
ice cream	40

6. How many fruit bowls and ice-cream treats were sold altogether? **100 items**
7. Which two items sold the least? **hot dogs & fries**
8. Which item was sold the most? **fruit bowls**
9. How many more hamburgers were sold than hot dogs? **20 more**
10. How many more chips were sold than fries? **20 more**

72   © Carson-Dellosa • CD-104628

Answer Key

Name _____ (3.MD.B.3)

Bar Graphs and Pictographs

Use the information in the charts to complete the graphs. Then, use the graphs to answer each question.

Grade Level	Pounds Donated
K	30
1st	40
2nd	40
3rd	60
4th	70
5th	20

Month	Number of Students
February	30
April	50
June	70
September	10

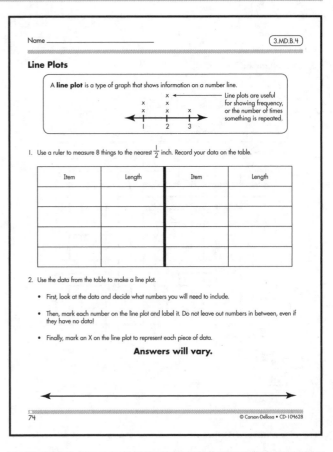

Lindy Elementary Food Drive

Each ☐ stands for 10 lb. of donated food.

Rocket Day Fun!

1. Which grade level had the most donations? **4th**

2. Which grade level had only 20 pounds of food donated? **5th**

3. What was the total amount of food donated for the entire school? **260 pounds**

4. How much more did fourth grade donate than fifth grade? **50 pounds**

5. Who donated more, first or third grade? **3rd**

6. How many more pounds did kindergarten donate than fifth grade? **10 pounds**

7. Which month had the greatest attendance at Rocket Day? **June**

8. How many more students attended Rocket Day in February than in September? **20 students**

9. Which month had 50 students attend? **April**

10. How many more students attended in June than in April? **20 students**

11. What was the total number of students that attended Rocket Day in all? **160 students**

12. Which month had the least number of students attending? **September**

© Carson-Dellosa • CD-104628   73

Name _____ (3.MD.B.4)

Line Plots

A **line plot** is a type of graph that shows information on a number line.

Line plots are useful for showing frequency, or the number of times something is repeated.

1. Use a ruler to measure 8 things to the nearest $\frac{1}{2}$ inch. Record your data on the table.

Item	Length	Item	Length

2. Use the data from the table to make a line plot.

- First, look at the data and decide what numbers you will need to include.

- Then, mark each number on the line plot and label it. Do not leave out numbers in between, even if they have no data!

- Finally, mark an X on the line plot to represent each piece of data.

Answers will vary.

74   © Carson-Dellosa • CD-104628

Name _____ (3.MD.B.4)

Line Plots

1. Use a ruler to measure 10 things to the nearest $\frac{1}{2}$ inch. Record your data on the table.

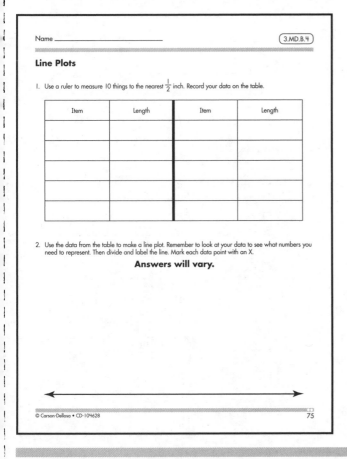

2. Use the data from the table to make a line plot. Remember to look at your data to see what numbers you need to represent. Then divide and label the line. Mark each data point with an X.

Answers will vary.

© Carson-Dellosa • CD-104628   75

Name _____ (3.MD.B.4)

Line Plots

1. Use a ruler to measure 10 things to the nearest $\frac{1}{4}$ inch. Record your data on the table.

Item	Length	Item	Length

2. Use the data from the table to make a line plot.

Answers will vary.

76   © Carson-Dellosa • CD-104628

Answer Key

3.MD.C.5, 3.MD.C.6

Name _____

Understanding Area

Area is the number of square units it takes to cover the surface of a figure.

To find area, count the number of squares it takes to cover the shape. The squares must touch along the edges with no overlap and no gaps.

Area is measured in *square units*, such as *square inches* or *square centimeters*. If the unit is not known, you can use square units as shown in the example.

Area (A) = 6 square units

Find the area of each figure.

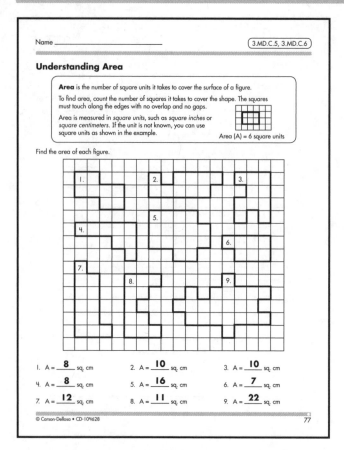

1. A = __8__ sq. cm   2. A = __10__ sq. cm   3. A = __10__ sq. cm
4. A = __8__ sq. cm   5. A = __16__ sq. cm   6. A = __7__ sq. cm
7. A = __12__ sq. cm   8. A = __11__ sq. cm   9. A = __22__ sq. cm

---

3.MD.C.5, 3.MD.C.6

Name _____

Understanding Area

Remember, **area** is the number of square units with no gaps or overlaps in a figure. When measuring area, it helps to mark each square as you count it.

Find the area of each figure.

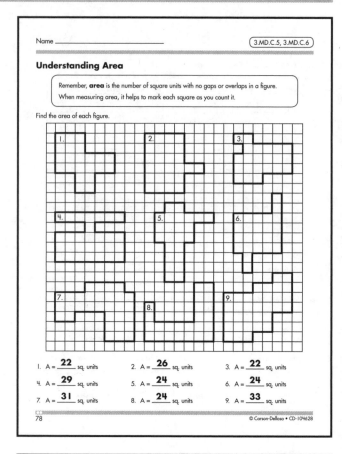

1. A = __22__ sq. units   2. A = __26__ sq. units   3. A = __22__ sq. units
4. A = __29__ sq. units   5. A = __24__ sq. units   6. A = __24__ sq. units
7. A = __31__ sq. units   8. A = __24__ sq. units   9. A = __33__ sq. units

---

3.MD.C.5, 3.MD.C.6

Name _____

Understanding Area

Remember, **area** is the number of square units with no gaps or overlaps in a figure. When measuring area, it helps to mark each square as you count it.

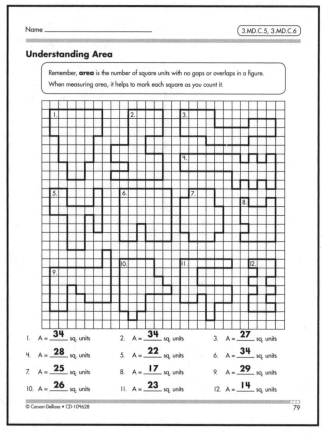

1. A = __34__ sq. units   2. A = __34__ sq. units   3. A = __27__ sq. units
4. A = __28__ sq. units   5. A = __22__ sq. units   6. A = __34__ sq. units
7. A = __25__ sq. units   8. A = __17__ sq. units   9. A = __29__ sq. units
10. A = __26__ sq. units   11. A = __23__ sq. units   12. A = __14__ sq. units

---

3.MD.C.5, 3.MD.C.6

Name _____

Finding Area

Area is the number of square units within a space. Area is measured in different units such as *square feet* or *square centimeters*. For example, there are 14 square units in this figure.

Area (A) = 14 square units

Find the area of each figure.

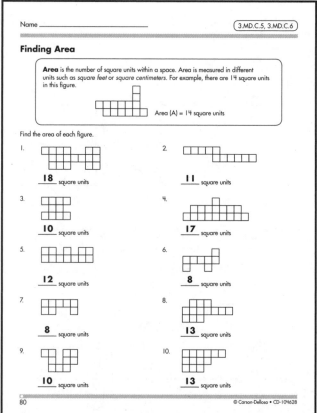

1. __18__ square units   2. __11__ square units
3. __10__ square units   4. __17__ square units
5. __12__ square units   6. __8__ square units
7. __8__ square units   8. __13__ square units
9. __10__ square units   10. __13__ square units

Answer Key

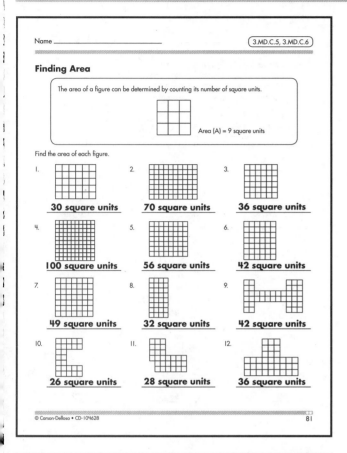

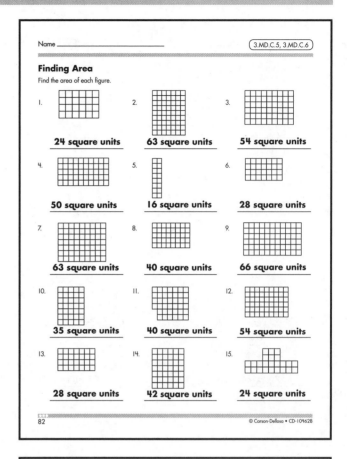

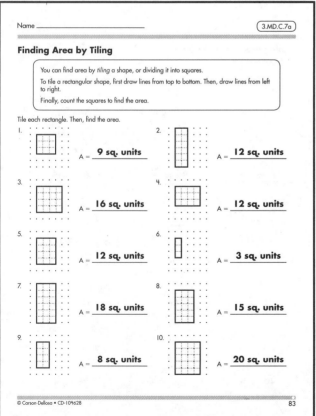

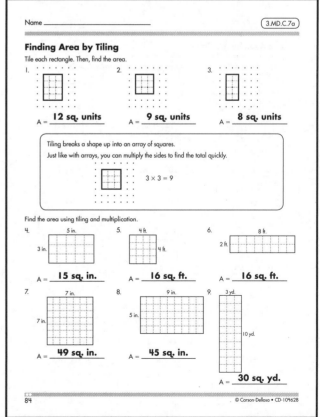

Answer Key

Name _____ 3.MD.C.7a

Finding Area by Tiling
Find the area of each rectangle.

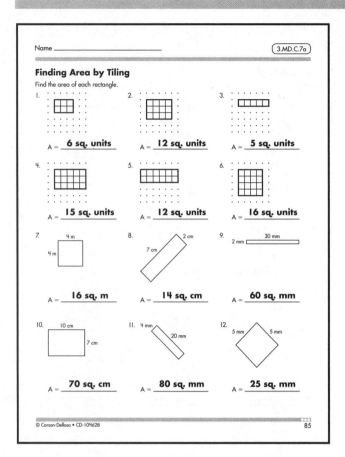

1. A = **6 sq. units**
2. A = **12 sq. units**
3. A = **5 sq. units**
4. A = **15 sq. units**
5. A = **12 sq. units**
6. A = **16 sq. units**
7. A = **16 sq. m**
8. A = **14 sq. cm**
9. A = **60 sq. mm**
10. A = **70 sq. cm**
11. A = **80 sq. mm**
12. A = **25 sq. mm**

Name _____ 3.MD.C.7a, 3.MD.C.7b

Finding Area by Multiplying

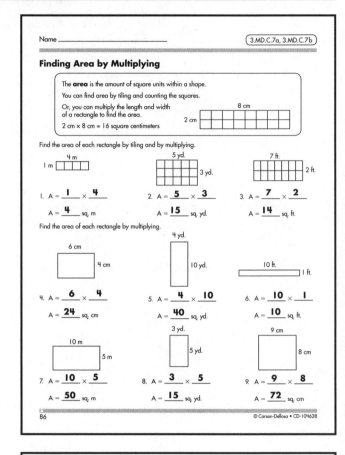

The **area** is the amount of square units within a shape.
You can find area by tiling and counting the squares.
Or, you can multiply the length and width of a rectangle to find the area.
2 cm × 8 cm = 16 square centimeters

Find the area of each rectangle by tiling and by multiplying.

1. A = **1** × **4**
   A = **4** sq. m
2. A = **5** × **3**
   A = **15** sq. yd.
3. A = **7** × **2**
   A = **14** sq. ft.

Find the area of each rectangle by multiplying.

4. A = **6** × **4**
   A = **24** sq. cm
5. A = **4** × **10**
   A = **40** sq. yd.
6. A = **10** × **1**
   A = **10** sq. ft.
7. A = **10** × **5**
   A = **50** sq. m
8. A = **3** × **5**
   A = **15** sq. yd.
9. A = **9** × **8**
   A = **72** sq. cm

Name _____ 3.MD.C.7a, 3.MD.C.7b

Finding Area by Multiplying

Remember, you can find the area of a rectangle by tiling, or multiplying the length and width.

Find the area of each rectangle by tiling and by multiplying.

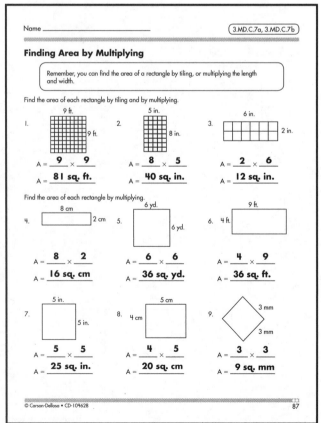

1. A = **9** × **9**
   A = **81** sq. ft.
2. A = **8** × **5**
   A = **40** sq. in.
3. A = **2** × **6**
   A = **12** sq. in.

Find the area of each rectangle by multiplying.

4. A = **8** × **2**
   A = **16** sq. cm
5. A = **6** × **6**
   A = **36** sq. yd.
6. A = **4** × **9**
   A = **36** sq. ft.
7. A = **5** × **5**
   A = **25** sq. in.
8. A = **4** × **5**
   A = **20** sq. cm
9. A = **3** × **3**
   A = **9** sq. mm

Name _____ 3.MD.C.7b

Finding Area by Multiplying
Find the area of each figure.

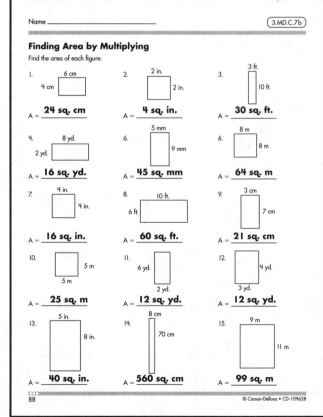

1. A = **24 sq. cm**
2. A = **4 sq. in.**
3. A = **30 sq. ft.**
4. A = **16 sq. yd.**
5. A = **45 sq. mm**
6. A = **64 sq. m**
7. A = **16 sq. in.**
8. A = **60 sq. ft.**
9. A = **21 sq. cm**
10. A = **25 sq. m**
11. A = **12 sq. yd.**
12. A = **12 sq. yd.**
13. A = **40 sq. in.**
14. A = **560 sq. cm**
15. A = **99 sq. m**

Answer Key

Finding Area of Rectilinear Figures

3.MD.C.7d

To find the area of complex rectangular figures, divide the figure into two or more rectangles.

Then, find the area of each rectangle.

Add the areas together to find the total area of the shape.

Find the area of each figure.

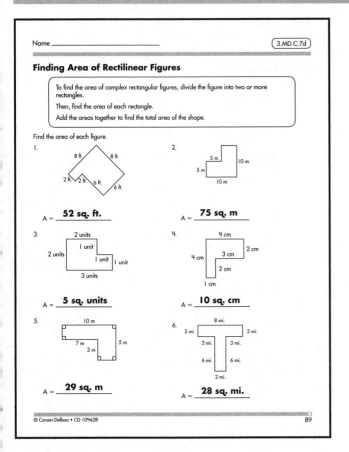

1. A = **52 sq. ft.**

2. A = **75 sq. m**

3. A = **5 sq. units**

4. A = **10 sq. cm**

5. A = **29 sq. m**

6. A = **28 sq. mi.**

© Carson-Dellosa • CD-104628   89

Finding Area of Rectilinear Figures

3.MD.C.7d

Remember, to find the area of complex rectangular figures,

1. Divide the figure into two or more rectangles.

2. Find the area of each rectangle.

3. Add the areas together to find the total area of the shape.

Find the area of each figure.

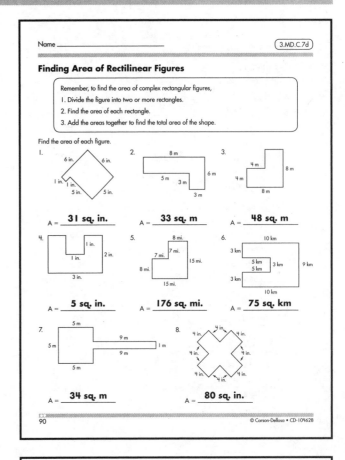

1. A = **31 sq. in.**

2. A = **33 sq. m**

3. A = **48 sq. m**

4. A = **5 sq. in.**

5. A = **176 sq. mi.**

6. A = **75 sq. km**

7. A = **34 sq. m**

8. A = **80 sq. in.**

90   © Carson-Dellosa • CD-104628

Finding Area of Rectilinear Figures

3.MD.C.7d

Find the area of each figure. All measurements are in centimeters.

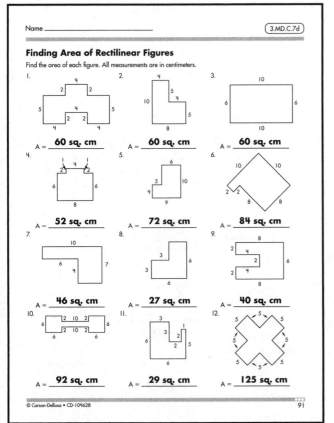

1. A = **60 sq. cm**

2. A = **60 sq. cm**

3. A = **60 sq. cm**

4. A = **52 sq. cm**

5. A = **72 sq. cm**

6. A = **84 sq. cm**

7. A = **46 sq. cm**

8. A = **27 sq. cm**

9. A = **40 sq. cm**

10. A = **92 sq. cm**

11. A = **29 sq. cm**

12. A = **125 sq. cm**

© Carson-Dellosa • CD-104628   91

Finding Perimeter

3.MD.D.8

Perimeter is the total distance around a given figure. To find the perimeter, add the lengths of the sides of the figure.

Example: P = perimeter

P = 4 cm + 8 cm + 4 cm + 8 cm

P = 24 cm

Find the perimeter of each figure.

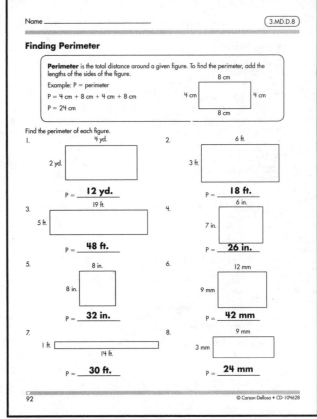

1. P = **12 yd.**

2. P = **18 ft.**

3. P = **48 ft.**

4. P = **26 in.**

5. P = **32 in.**

6. P = **42 mm**

7. P = **30 ft.**

8. P = **24 mm**

92   © Carson-Dellosa • CD-104628

Answer Key

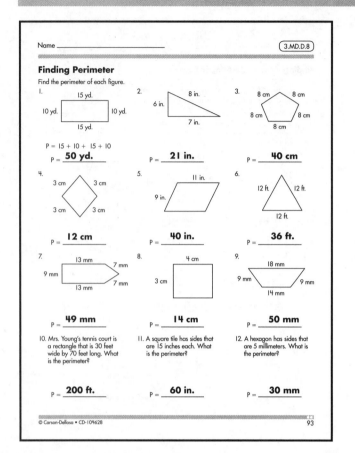

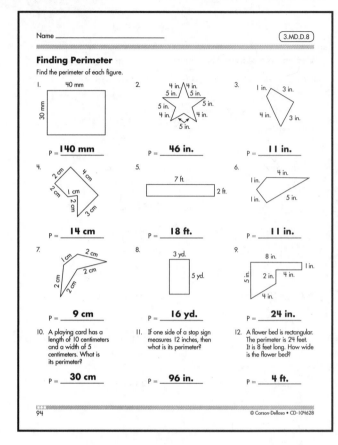

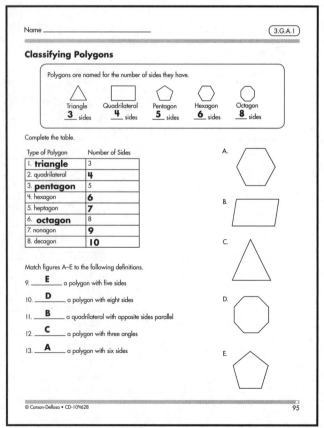

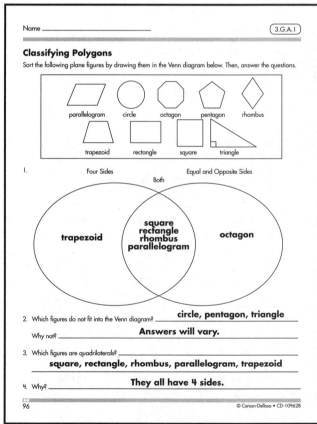

Answer Key

(3.G.A.1)

Classifying Polygons

Complete each statement with *All, Some, No,* or *None.*

1. __All__ rectangles have 4 vertices. __All__ rectangles are parallelograms.
   __No__ rectangles are circles.

2. __All__ quadrilaterals have 4 sides. __Some__ quadrilaterals are rectangles.
   __No__ quadrilaterals have just 3 vertices.

3. __Some__ polygons are quadrilaterals. __Some__ polygons have equal sides.
   __No__ polygons have curved sides.

4. __All__ pentagons have 5 sides. __All__ hexagons have more than 5 sides.
   __No__ octagons have less than 5 sides.

5. __No__ triangles have 4 vertices. __No__ triangles are quadrilaterals.
   __All__ triangles have 3 vertices.

Look at the diagrams to complete the statements using *All, Some, No,* or *None.*

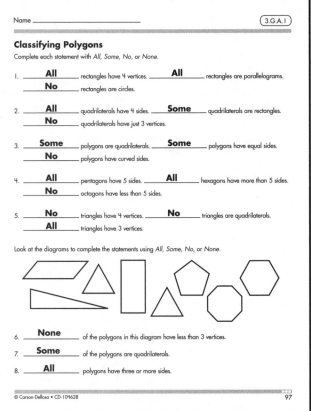

6. __None__ of the polygons in this diagram have less than 3 vertices.

7. __Some__ of the polygons are quadrilaterals.

8. __All__ polygons have three or more sides.

© Carson-Dellosa • CD-104628   97

(3.MD.B.3, 3.G.A.1)

Recognizing Polygons

Look at the shapes on the right. Follow the directions.

1. Color the triangles green.
2. Color the quadrilaterals red.
3. Color the pentagons blue.
4. Color the hexagons orange.
5. Color the octagons purple.

Check students' coloring.

6. Use the shapes to make a chart.

Shape	Number Found
triangle	9
quadrilateral	8
pentagon	8
hexagon	7
octagon	3

7. Color one box on the graph for each shape.

Number of Shapes (y-axis: 0–20)

Less Than 4 Sides	Quadrilateral	More Than 4 Sides
9	8	18

98   © Carson-Dellosa • CD-104628

(3.G.A.1)

Recognizing Polygons

Identify each type of polygon as a triangle, quadrilateral, or pentagon.

1. __quadrilateral__   2. __triangle__   3. __triangle__   4. __pentagon__

5. __quadrilateral__   6. __triangle__   7. __pentagon__   8. __pentagon__

9. __quadrilateral__   10. __triangle__   11. __quadrilateral__   12. __pentagon__

> A parallelogram is a special type of quadrilateral that has opposite sides that are parallel and the same length.   _parallel_
> A rectangle is a parallelogram that has four right angles.   A square is a rectangle with four sides equal in length.

Identify each type of polygon as a parallelogram, rectangle, or square.

13. __square__   14. __parallelogram__

15. __rectangle__   16. __parallelogram__

© Carson-Dellosa • CD-104628   99

(3.G.A.1)

Recognizing Polygons

A **quadrilateral** is a closed figure with four sides and four angles. Make four different quadrilaterals. Record your figures here. **Drawings will vary.**

1.   2.   3.   4.

What makes each of these figures a quadrilateral? **They have four sides and four angles.**

There are special types of quadrilaterals.
• A **trapezoid** is a quadrilateral with just one set of parallel sides.
• A **parallelogram** is a quadrilateral with two sets of parallel sides.
• A **rectangle** is a parallelogram with four right angles.
• A **square** is a rectangle with four sides of equal length.

Make the figures described. Record them on the grids below. **Drawings will vary.**

5. a trapezoid   6. a parallelogram   7. a rectangle   8. a square

9. a parallelogram that isn't a rectangle   10. a parallelogram that is a square   11. a quadrilateral that is a trapezoid   12. a rectangle that is a square

13. a rectangle that isn't a square   14. a quadrilateral that isn't a trapezoid or parallelogram   15. a parallelogram that is a rectangle   16. a square that is a rectangle

100   © Carson-Dellosa • CD-104628

Answer Key

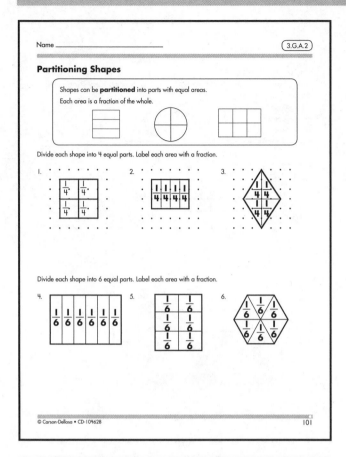

Name _____ 3.G.A.2

Partitioning Shapes

Shapes can be **partitioned** into parts with equal areas.
Each area is a fraction of the whole.

Divide each shape into 4 equal parts. Label each area with a fraction.

Divide each shape into 6 equal parts. Label each area with a fraction.

© Carson-Dellosa • CD-104628   101

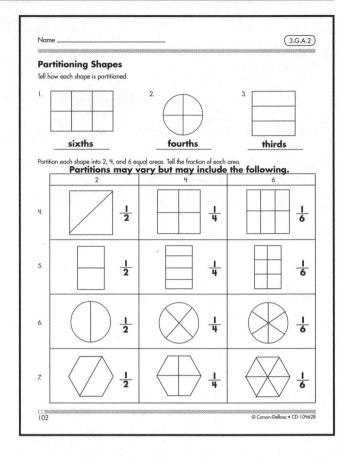

Name _____ 3.G.A.2

Partitioning Shapes

Tell how each shape is partitioned.

1. _____sixths_____   2. _____fourths_____   3. _____thirds_____

Partition each shape into 2, 4, and 6 equal areas. Tell the fraction of each area.

Partitions may vary but may include the following.

	2		4		6	
4.		$\frac{1}{2}$		$\frac{1}{4}$		$\frac{1}{6}$
5.		$\frac{1}{2}$		$\frac{1}{4}$		$\frac{1}{6}$
6.		$\frac{1}{2}$		$\frac{1}{4}$		$\frac{1}{6}$
7.		$\frac{1}{2}$		$\frac{1}{4}$		$\frac{1}{6}$

102   © Carson-Dellosa • CD-104628

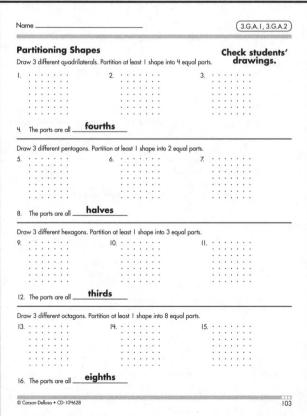

Name _____ 3.G.A.1, 3.G.A.2

Partitioning Shapes

Check students' drawings.

Draw 3 different quadrilaterals. Partition at least 1 shape into 4 equal parts.

1. ·······   2. ·······   3. ·······

4. The parts are all _____fourths_____

Draw 3 different pentagons. Partition at least 1 shape into 2 equal parts.

5. ·······   6. ·······   7. ·······

8. The parts are all _____halves_____

Draw 3 different hexagons. Partition at least 1 shape into 3 equal parts.

9. ·······   10. ·······   11. ·······

12. The parts are all _____thirds_____

Draw 3 different octagons. Partition at least 1 shape into 8 equal parts.

13. ·······   14. ·······   15. ·······

16. The parts are all _____eighths_____

© Carson-Dellosa • CD-104628   103